OBSERVING COMPLEXITY

OBSERVING COMPLEXITY

SYSTEMS THEORY AND POSTMODERNITY

William Rasch and Cary Wolfe, Editors

University of Minnesota Press / Minneapolis London

Copyright 2000 by the Regents of the University of Minnesota

Published by the University of Minnesota Press
111 Third Avenue South, Suite 290
Minneapolis, MN 55401-2520
http://www.upress.umn.edu

Library of Congress Cataloging-in-Publication Data

Observing complexity : systems theory and postmodernity / William Rasch and
    Cary Wolfe, editors.
        p.   cm.
    Includes bibliographical references and index.
    ISBN 0-8166-3297-9 (alk. paper) — ISBN 0-8166-3298-7 (pbk. : alk. paper)
    1. Postmodernism.   2. Systems theory.   I. Rasch, William, 1949–   II. Wolfe,
    Cary.
    B831.2 .027 2000
    003—dc21   99-050623

Printed in the United States of America on acid-free paper

The University of Minnesota is an equal-opportunity educator and employer.

11 10 09 08 07 06 05 04 03 02 01 00   10 9 8 7 6 5 4 3 2 1

Contents

Acknowledgments

Much of the work presented in this collection arose as part of a two-year project, involving many hands, which included a special issue of *New German Critique* 61 (winter 1994) on Niklas Luhmann edited by William Rasch and Eva Knodt; a conference titled "Systems Theory and the Postmodern Condition" held at Indiana University in September 1994, organized by Rasch and Knodt; a yearlong multidisciplinary faculty seminar on the history of cybernetics and systems theory at Indiana University in 1994, conducted by Cary Wolfe and William Rasch; and, most important, two special issues of *Cultural Critique,* 30 and 31 (spring and fall 1995) in which most of these essays originally appeared.

We would like to thank the following for contributing, directly or indirectly, to this project: the Dean of Faculties office at Indiana University, Bloomington, for supporting the conference, the multidisciplinary seminar, and the *Cultural Critique* special issues; the German Academic Exchange Service, the Fritz Thyssen Stiftung, and various schools, departments, and institutes of Indiana University for supporting the conference; Skip Willman, Anita Schaad, and Tina O'Donnell for help in preparing the manuscripts; and Lily Kay and Tom Cohen, who read this project for the University of Minnesota Press, for their insightful comments and helpful suggestions.

We owe a special debt of gratitude to Donna Przybylowicz and Abdul R. JanMohamed for allowing us to present this project originally in the pages of *Cultural Critique.*

Introduction: Systems Theory and the Politics of Postmodernity

William Rasch and Cary Wolfe

Dilemmas of Enlightenment; or, What Is to Be Done?

In 1944 Max Horkheimer and Theodor W. Adorno attempted to come to terms with the disaster that had befallen mid-century Europe by posing a seemingly insoluble problem. "The dilemma that faced us in our work," they wrote in their introduction to *Dialectic of Enlightenment,* "proved to be the first phenomenon for investigation: the self-destruction of the Enlightenment." The dramatic self-immolation of Western culture, represented for them not only by the brutalities of German fascism but also by the relentless instrumentalism of both Soviet Marxism and American consumerism, did not, they assured their readers, turn them into enemies of reason. On the contrary, "We are wholly convinced—and therein lies our *petitio principii*—that social freedom is inseparable from enlightenment thought." "Nevertheless," they continued, "we believe that we have just as clearly recognized that the notion of this very way of thinking, no less than the actual historic forms—the social institutions—with which it is interwoven, already contains the seed of the reversal universally apparent today." For these two champions *and* critics of the Enlightenment tradition, belief in the moral autonomy of the subject and in the simultaneous material and spiritual progress of the social collective served, to be sure, as the basis for faith in progressive political movements. But these basic tenets of autonomy and perfectibility might also engender some striking unintended consequences of the sort often associated with the "technocratic" and the "pragmatic." The task they set themselves, then, was clear: to subject Enlightenment thought to a rigorous self-examination. For "If consideration of the destructive aspect of progress is left to its enemies, blindly pragmatized thought loses its transcending quality and its relation to truth" (xiii).

It is this last claim, more than any other, that defines both the nature and the failure of the project of Horkheimer and Adorno. "Blindly pragmatized thought," their name for modern, mathematized, instrumental rationality, would appear to have the capacity for both positive and negative effects, but the ability to distinguish between the politically progressive and emancipatory, on the one hand, and the horrors of the "administered," "managerial," and "technocratic" society, on the other, requires the ability not only to know "truth" when one sees it, but also to specify the relation between that truth and the practical forms of social and material practice that embody it. For without this privileged relation to an absolute standard of truth, "blindly pragmatized thought" would not be able to transcend its own immanence, its own blind self-replication. In their introduction, then, Horkheimer and Adorno maintain the time-honored view that without an assured access to truth, critique has no basis and is thus powerless. And hence, the "failure" of their project resides in the fact that their analysis of the concept of Enlightenment is also, in a fundamental sense, a reproduction of its constitutive dilemma; for in their analyses of the culture industry, anti-Semitism, and much else besides, this relation between critique and truth is everywhere presupposed, but nowhere demonstrated or brought to light. What Horkheimer and Adorno both reveal and enact, then, is an Enlightenment tradition that performs as if it were philosophically grounded, rooted in more than just emancipatory desire, only to find itself, like a startled cartoon character, suspended in midair, miles above the bedrock it assumes is right under its feet. From this vantage, the supposedly philosophical and critical gesture of Enlightenment reduces instead to an ethos, a personal display of political commitment or ethical purity that, in the end, foregoes rational argument and theoretical rigor for moral persuasion and the inducement of compassion and sympathy. If this is the case, then the politically progressive Enlightenment tradition must be viewed less as a philosophically grounded political program and more as itself a pragmatic set of solutions and proposals, thoroughly historical in character, that appeal for their privileged status upon the spreading of a shared personal ethic, a kind of latter-day monastic order that collects alms for the poor by dispensing guilt and the means for absolving that guilt.

Over the course of the last half century, leftist thought in both the United States and Europe has wrestled with this legacy, for what Horkheimer and Adorno describe is the widely felt disjuncture between the epistemological antifoundationalism that emerges from a variety of twentieth-century philosophical movements and traditions—especially the various hermeneutical, poststructuralist, and pragmatist philosophies associated with names like Wittgenstein, Heidegger, Gadamer, Derrida, and Rorty—and the political commitments that many intellectuals on the left have wanted to maintain. While these political commitments and the forms of social and cultural theory they have underwritten have typically relied on a representationalist view of the relationship between knowledge and practice (a point to which we will return in a moment), the great modernist philosophical paradigms upon which these social theories were based have been subject to a relentless battering since the collapse of the Hegelian and neo-Kantian projects of the nineteenth century, throwing into permanent question the "subjective" or "objective" foundations of both knowledge and action. After Nietzsche's perspectivism, Max Weber's differentiation of value spheres, and Wittgenstein's plurality of language games, the problem becomes not deciding which perspective, which value, or which language game is the foundation for all the others, but rather how one is to live in a truly pluralist world, a world in which a multiplicity of values and rationalities compete on an equal footing, each asserting equal "rights," and none, ultimately, being able to trump the other. In such a differentiated world, the link between "blindly pragmatized thought" and "truth" becomes more or less permanently problematic, because instrumental rationality and metaphysical truth emerge as competing values, each relativized, and each observing the other, as it were, from the outside.

One of the more interesting developments on the intellectual landscape of the past thirty years, we think, is that the distinction between *accepting* and *rejecting* this "relativist" or "pluralist" or "differentiated" view of modernity (and beyond that, of course, its intensification under "postmodernity") does not correspond in any neat way with the traditional political distinction between "left" and "right." For many on the right *and* many on the left, the acceptance of differentiated modernity and its concomitant relativization of truth claims is seen as nothing other than pyrrhonic skepticism or out-and-out nihilism, whereby supposedly progressive or radical poststructuralists such as Jacques Derrida or Michel Foucault suddenly find themselves aligned, by their more traditional friends on the left, with conservatism itself (as in, for example, Jürgen Habermas's characterization, in his *Philosophical Discourse of Modernity*, of Foucault as a "young conservative"). In responding to this "moral crisis," conservatives have it easier, of course, for they can turn (or return) to

God, the pursuit of truth, the need for a civic religion, and ultimately to faith in the unity of the Good, the Beautiful, and the True. Since the noble search for such truth and unity (however delayed and deferred they always prove to be) is part of the nineteenth-century liberal tradition itself from which we inherit most of our political and civic institutions and habits, the need for arguing this position to the general public is greatly reduced. As Leo Strauss, the intellectual force behind much of contemporary neoconservatism, puts it:

> All political action is . . . guided by some thought of better and worse. But thought of better or worse implies thought of the good. The awareness of the good which guides all our actions has the character of opinion: it is no longer questioned but, on reflection, it proves to be questionable. The very fact that we can question it directs us towards such a thought of the good as is no longer questionable—towards a thought which is no longer opinion but knowledge. All political action has then in itself a directedness towards knowledge of the good: of the good life, or of the good society. For the good society is the complete political good. (10)

To deny this compelling chain of reasoning would seem to be unreasonable, even willfully evil. Indeed, the conservative political retrenchment we have witnessed over the past two or three decades may be viewed as a reaction to a seemingly casual and reckless acceptance, if not embrace, of the very uncertainty that is produced by the proliferation of competing values, bodies of specialized knowledge, and the lack of a standard— either intellectual or moral—by which this competition can be uniformly resolved. The conservative revival, in other words, gains its strength by offering the certainty of supposedly traditional values to those for whom the contingency of modernity is disquieting. Accordingly, moral exhortation, disguising itself as political philosophy, replaces real concern with, and rigorous examination of, the nature of the political in a historical context of (post)modernity that cannot so easily have its contingency wished away.

In confronting this dilemma, the left does not have it so easy. Because the Hegelian-Marxist synthesis has long since lost its credibility (except, perhaps, for the remarkable Fredric Jameson!), there can be no turning back to the proletariat or, after that, the Third World, as historical agents who would lead us all into the promised land in which "productive property is administered in the general interest not just out of 'good intentions' but with rational necessity," and in which "the life of the whole and the individuals alike is produced not merely as a natural effect but as the consequence of rational designs that take account of the happiness of individuals in equal measure" (Horkheimer 28, 29). This is not to say, of course, that these may not be desirable ends, but only that their desirability is pragmatic and not foundational, not grounded in "rational necessity" or the dialectically necessary emergence of a particular Subject of History.

The response to this state of affairs by what the press calls the "multi-cultural left" has usually been little more than polemical and occasional, a distant, fractured echo of an earlier self-confidence that felt itself to be unassailably grounded in what Jean-François Lyotard has famously called the "meta-narratives of emancipation" embodied by the American, French, and Russian Revolutions (and beyond that, the cultural revolution of Maoism, the toppling of Batista's Cuba by Castro, the anticolonial wars of liberation in Asia and Africa, and the mass movements of the late 1960s in both Europe and the United States). The reasons for this diminished leftism are not hard to find, and they are, ironically, rooted in the very theoretical and intellectual accomplishments of the left itself over the past thirty years. In observing the advances that have been made in what are called the "new social movements" (gay and lesbian rights, feminism, environmentalism, and animal rights, just to name a few), it seems clear that they have in practice continued to trade upon the traditional narratives of "rights" and "freedoms" grounded in the Enlightenment paradigm of the autonomous, rational individual with its capacities, potentialities, and interests, busily actualizing its freedom to "life, liberty, and the pursuit of happiness." At the same time, however, this very philosophical and political tradition—let's call it Enlightenment humanism for the moment—has long since been thrown into radical question as a *theoretical* foundation for politics and the subject's relationship to it, not only by salient critiques within the Marxist tradition itself (C. B. McPherson, Fredric Jameson, Louis Althusser and his followers) and in psychoanalysis (Lacan and his epigones, such as Zizek), but also, and most notably, by the poststructuralist lineage of Foucault, Lyotard, Derrida, Baudrillard, and Deleuze and Guattari. What we find, in other words, is a rather dramatic disjuncture between the rhetoric and ruling paradigms of political praxis in the new social movements, and the theoretical and epistemological commitments of most of the leading intellectuals who see themselves aligned with those movements.

To put it another way—taking very briefly the examples of Marxism, feminism, and liberalism respectively (if only schematically)—it is not that the phenomenological and linguistic calcification that an older Marxist vocabulary called "reification" does not exist; it is rather that the *political* critique of reification traditionally made available by reference to the *theoretical* science/ideology distinction is no longer persuasive or tenable, not only in light of the deconstructive critique of logocentrism, but also because of important developments within the twentieth-century philosophy of science itself (some of them discussed in these pages). Similarly, it seems obvious that patriarchal oppression under the gender system remains intact; but attempts to theoretically ground feminist political solidarity against patriarchal oppression by reference to the ontological givenness of "women's experience" seem dubious at best in light of the problematization of that notion, and of the category of gender in general, by feminists like Judith Butler and work in "queer theory." Finally, within liberalism, the Habermasian strategy of welding democratic consensus out of the proliferation of democratic differences in subject positions, discourses, and what Lyotard calls "differends" seems less and less persuasive, as even fellow liberals like Richard Rorty have pointed out. For insofar as *political* consensus depends upon a *theoretical* appeal to an Enlightenment metanarrative of rationality to adjudicate the proliferation of language games (whereby most of postmodern theory ends up being condemned as "nihilistic" and "neoconservative"), and insofar as that appeal relies upon the presumption of a utopian "ideal speech situation," it presumes the very consensus that needs to be demonstrated and achieved. But if there is no consensus affirming Habermas's consensual theory of truth, then how can he continue to insist on the truth of his theory without committing the very same performative contradiction he finds so loathsome in others?

It is clear, then, that we on the left have lost both our political confidence and our political effectivity in no small part because we have brilliantly gone about the business of removing the theoretical ground from under our feet, and we have *not* been able to turn this loss—despite being blamed for it—into our gain. But this quintessentially modern ability to turn a loss of ground into a political gain—*this,* as a contemporary, ironically inclined Lenin might say, is precisely what is to be done.

Systems Theory and the System of Theory

To do it, however, we have to widen our frame of reference. We feel that the humanities are now experiencing the diminishing returns of an increasingly unproductive quarrel about the consequences of the "relativism" that seems unavoidable once it is recognized that no necessary link between rationality and truth exists. It is not too much to say that this problem has been *the* central conundrum for politically attuned intellectuals in the second half of this century, especially during the past two or three decades. On the one hand, the *epistemological* critique of realism and positivism on behalf of the contingency and constructedness of knowledge has seemed to hold great *political* promise, because it offers hope that a world that is contingently constructed can also be *differently* constructed, i.e., more *justly* constructed. But on the other hand, that very constructivist critique, which opened up so many political possibilities, has also proved to be a liability for those seeking theoretical grounds for political practice. Having problematized the epistemological claims undergirding the privileges of those in political power, leftist intellectuals find their own claims and their own authority in jeopardy as well. As a result, they often find themselves driven back upon representationalist notions jarringly at odds with their most exhilarating, pathbreaking, and rigorous work. The challenge we face, then, is not the taking of this or that side within a representationalist account of the relationship between theory and political practice, but rather the continued questioning of the representationalist frame itself.

Fortunately, developments across a whole range of disciplines over the past thirty years have enabled, indeed demanded, such a long-overdue reassessment. In recent decades, theoretical interest—in philosophical pragmatism and the various poststructuralisms, in post-Kuhnian philosophy and sociology of science, and, more recently, in complexity, chaos, and systems theories—has time and again returned to fundamental epistemological problems, problems originally "solved" by Kant and the Enlightenment only to be raised again by waves of successors. Though the "relativization" of epistemological certitude under pressure from these theoretical developments was originally welcomed by many in the '60s and '70s as politically emancipatory, during the '80s initial reservations grew into a crescendo of critique, from outright rejection of the "neoconservative nihilism" of poststructuralists such as Foucault and de Man, to more thoughtful responses to the challenge of "philosophical skepticism." The reactions on the left to the epistemological crisis associated with postmodernity (but firmly rooted, as Niklas Luhmann argues, in modernity) have been manifold, including (just to name a few) the attempted maintenance or reconstruction of historical materialism (as in the work of Roy Bhaskar or Perry Anderson) and of the concept of objectivity (as in feminist philosophers of science like Sandra Harding), the new historicist search for contextual embeddedness and "thick description," and a renewed debate about the centrality of ethics (whether of the

Habermasian variety or fueled by the identity politics of marginalized others) over and against politics as such.

As important as these debates are, we nevertheless see in them, as we have already suggested, a certain characteristic evasion rather than investigation of crucial epistemological problems, an *evasion* that often leads to the avoidance or repression of central and intractable theoretical problems in the name of political expediency. We feel, however, that an important and in some sense distinguishing legacy of twentieth century thought—the so-called "loss of reference" with its concomitant dislocation of epistemological security—cannot profitably be glossed over, not just because it presents an interesting intellectual challenge, but because it presents an even more important political one. What we wish to foreground in this collection of essays and interviews, therefore, is *not* some new (rationalist or, even, pragmatic) overcoming of the tension between knowledge and action first made palpably manifest in Kant's critical and political writings; nor is it the elegant sidestepping of the problem by appeals to the sovereignty of aesthetic or reflective judgment, as in Lyotard's neo-Kantian championing of the sublime. On the contrary, we wish to focus on the inevitability of contingency that this unresolved tension cannot help but create, not as some new form of cultural pessimism or trendy nihilism, but rather as an appreciation of the fact that the causal disjunction of system and environment, discourse and its object, language and its referent, should be viewed not as a foreclosure of the possibility of informed political practice and social theory, but rather as an opening and an invitation to a philosophical and political pluralism, one whose commitment to democratic difference may be gauged precisely by the extent to which it squarely faces the loss of the referent and the contingency and materiality of knowledge and the practices that constitute it. Politically as well as epistemologically, the only way out is through.

We will return to the question of politics in the concluding section of our introduction, but for now we want to note that in arguing for the importance of this theoretical and historical orientation, we are in many ways simply responding belatedly to important theoretical developments outside the human and social sciences. In recent decades, the natural sciences have experienced (to use Thomas Kuhn's well-known phrase) a fundamental "paradigm shift" marked by an increased interest in the workings of complex, as opposed to simple, systems. As Warren Weaver put it in 1949, science up to the end of the nineteenth century succeeded in solving problems of simple systems (e.g., the movements of the solar system), while the science of the first half of the twentieth century learned, by means of statistical analysis and probability theory, to deal with problems of disorganized complexity. It was his contention that the task of science for the latter half of the twentieth century was to develop means of investigating the dynamics of an *organized* complexity not characterized by random behavior and therefore not explicable by the rules of probability.

As the spectacular advances of his day (in quantum physics, cybernetics, and information theory, for example) and of ours (most popularly represented by chaos theory) demonstrate, Weaver's prediction has been amply substantiated. One of the rubrics used to bring together these disparate conceptual models and approaches from a number of fields (e.g., biology, communications, mathematics, physics) has been the name "systems theory." Since the famous Macy conferences of the late 1940s and early 1950s, the amalgamation of approaches associated with systems theory—cybernetics, information theory, cognitive science, and related fields—has claimed to be *the* model for any truly unified transdisciplinary research project of the future. By calling traditional scientific concepts of linear causality, determinism, and reductionism into question and replacing (or at least supplementing) them with notions of circular causality, self-organization, indeterminacy, and the unpredictable emergence of order from disorder, systems theory worked toward developing a unified theory and methodological approach to investigate not just the classical simplicities of the mechanistically structured material world, but also the complexities of biological, cognitive, and even social systems. This shift in emphasis has enabled the sciences to evaluate the relativization (if not the breakdown) of the Newtonian worldview not as a crisis (as it was viewed at the beginning of this century), but rather as liberating, enabling more nuanced and comprehensive, yet still mathematically precise, descriptions of complex, nonlinear, dynamic phenomena. More importantly for our purposes, it has also enabled philosophers of science to theorize with great rigor and precision epistemological problems that have traditionally bedeviled philosophers, literary critics, and anyone concerned with interpretation and the problem of knowledge: chiefly, how to acknowledge the contingency and constructedness of all description and interpretation without at the same time falling into the trap of "anything goes" relativism.

Our goal in this volume, then, is to encourage the long-overdue reassessment of humanistic knowledge in light of these new developments in the sciences that can be encompassed by the general theoretical framework of systems theory. It is our belief that scholars in the humanities and sciences alike who are concerned with the general epistemological problems of description, observation, and interpretation have much to teach each other as the search for a truly transdisciplinary research program becomes an ever more pressing agenda at the beginning of the twenty-first century. Now more than ever, scholars in all fields need to be able to theorize the connections between different kinds of knowledge in different fields, to be able to situate their work in a world of tightly interlaced and "hybrid" networks (to use Bruno Latour's term) of human, technological, organic, and informational systems.

Within our own historical horizon, the search for an interdisciplinary theoretical framework stretches back to the "unity of science" movement of the 1930s and 1940s—a movement, however, which remained grounded in logical positivism and a traditional, mathematics-based, deterministic epistemology, where only logically precise, quantifiable knowledge counted *as* knowledge. Indeed, the entire point of scientific method from this perspective was to obviate the need for general epistemological questioning of the descriptive procedures of science and how they might be analyzed using the tools of philosophy, textual and cultural analysis, sociology, and other critical tools. But with the stunning reception, both popular and philosophical, of the science of quantum physics and its startling epistemological implications for the nature of observation suggested by Bohr and Heisenberg, the clear distinction between the scientific observer and the observed phenomenon was fundamentally called into question—and so were some of the more important assumptions about scientific method and the relationship between observer and observed.

In time, the general phenomenon of circularity—dramatized, as it were, by the demonstrable impossibility of cleanly distinguishing between observed and observing system on the subatomic level—would serve as the abiding concern for the scientific movement that is the most important precursor to our immediate context—the so-called cybernetics movement of the late 1940s and early 1950s. As Steve Joshua Heims explains in his social history of the movement, *Constructing a Social Science for Postwar America: The Cybernetics Group 1946–1953*, researchers such as Norbert Wiener, Julien Bigelow, Warren McCulloch, Walter Pitts, and John von Neumann were intensely interested in both the formal dynamics and practical implications of how living and mechanical systems alike depend upon circularity—the processing of negative and positive feedback, for instance—for maintenance, adaptation, and steering.

While interested in the epistemological and philosophical implications of circularity, recursivity, and the like, the group's work was also intensely pragmatic; it was, as Norbert Wiener put it, "the science of control and communication in the animal and machine." It attempted to apply mathematical procedures to everyday activities that science, for the most part, had avoided—activities of purposive action, steering, and homeostatic regulation in animals and machines, such as reaching for a glass of water, steering a ship with a rudder, and, most famously, the regulation of room temperature by means of a thermostat—all prime examples of the famous "negative feedback loop" in which information is processed by the system in such a way as to maintain the harmony or homeostasis of the system. In the total systemic loop of thermostat/room/furnace, for example, the incoming information from the room (variations from a set baseline in temperature) is processed in such a way (engaging the compensatory mechanism of the furnace) as to maintain the homeostasis of the system (the desired room temperature). Accordingly, the incoming information *seems* to cause the stabilization of the system, but in fact what makes those differences *information*—what makes them "differences that *make* a difference," to use Gregory Bateson's definition of information (Bateson 272)—is that the thermostat is set in such a way as to specify the differences relevant—and the sensitivity with which differences are recognized as relevant—to the system. Information, that is, is not a natural category or quantity "out there" in the environment, but is instead thoroughly code-specific. In this sense, the "cause" of the system's homeostasis is, in fact, the system itself. Hence the system is characterized by "recursivity," which, as Niklas Luhmann defines it, is a process which "uses the results of its own operations as the basis for further operations . . . its own outputs as inputs" ("Cognitive Program" 72).

What is interesting about this example, from both an engineering and an epistemological perspective, is how it highlights the way the circularity of the negative feedback loop rests upon a paradox. As Heims notes:

> In traditional thinking since the ancient Greeks a cause A results in an effect B. With circular causality A and B are mutually cause and effect of each other. Moreover, not only does A affect B but through B acts back on itself. The circular causality concept seemed appropriate for much in the human sciences. It meant that A cannot do things to B without being itself effected. (23)

If it seemed that with the classical sciences of simplicity, "many scientists came to act as if they believed that the world accommodated to what their mathematics could handle and ignored the rest," then the cyberneticians seemed willing, with the help of mathematics and engineering, to follow complexity wherever it would lead, out into the surprisingly complicated world we experience every day (Heims 16).

No matter how fascinating the research, and no matter how revolutionary its effects, however, cybernetics has traditionally been politically suspect on the left, not least because of its emphasis on understanding the formal dynamics of control and steering. Despite renegade figures like Gregory Bateson, who became something of a cult figure during the hippie '60s and the New Age '80s, the intense interest in homeostasis, management, and social planning in cybernetics was typically seen (as in Lilienfeld) as an ideological manifestation of midcentury "technoscience" and capitalism. It is against this background that both the epistemological and ethical concerns of the second generation of cybernetics researchers can be understood, for if first-wave cybernetics focuses primarily upon the capacity of circular causality to generate stability and systemic equilibrium, second-wave cybernetics (as embodied in the work of biologists Humberto Maturana and Francisco Varela, physicists and mathematicians Ilya Prigogine and Manfred Eigen, and social scientists such as Niklas Luhmann) emphasizes instead how recursivity can lead to quite unexpected systemic effects and to the unpredictable evolution of complex systems.

Even more important for our purpose is the theoretical pressure that second-order cybernetics exerts upon the relationship between the paradox of circular causality (or recursivity) and the contingency of all observations and interpretations. If the unidirectional chain of causality turns back on itself, such that A both affects B and is affected by its effects on B, then the clear-cut distinction between observing and observed system is thoroughly problematized. This relativization of observation, first brought into focus by the revolution in early twentieth-century physics, is the central concern of the colorful figure Heinz von Foerster, who—despite his *Whole Earth Catalog* ethos, aphoristic writing style, and goofy illustrations—is an important figure in the transition from first- to second-order cybernetics, attempting to draw out epistemological conclusions worthy of his more famous uncle, Ludwig Wittgenstein. Second-order cybernetics, he argues, is characterized by a sort of full disclosure, as it were, of the problem of observation, how it is always already in a circular relation to that which is observed:

> (i) Observations are not absolute but relative to the observer's point of view (i.e., his coordinate system: Einstein); (ii) Observations affect the observed so as to obliterate the observer's hope for prediction (i.e., his uncertainty is absolute: Heisenberg).

After this, we are now in the possession of the truism that a description (of the universe) implies the one who describes (observes it). (von Foerster 258)

We can see in the work of Maturana and Varela how the treatment of the problem of observation by second-order cybernetics attempts to move beyond the traditional philosophical impasse of realism versus idealism, which has consistently hamstrung any attempt to do justice to the contingency of observation and interpretation without, at the same time, raising the feared specter of relativism. As they put it, "If we deny the objectivity of a knowable world, are we not in the chaos of total arbitrariness because everything is possible?" The way to "cut this apparent Gordian knot," they write, is to realize that the first principle of any sort of knowledge whatsoever is that "everything said is said by someone"— to foreground, in short, the problem of observation (135). But to theorize our observation of *that* observation, we need to realize the fact that "everything that is said" is said by means of fundamental distinctions (between x and not-x, let's say) that are, according to George Spencer Brown in *Laws of Form,* either paradoxical or tautological. As Ranulph Glanville and Francisco Varela remind us, "in the case of the elementary the last distinction in intension—we require that its distinction has no inside and, at the same time, we place in this non-existent inside a further distinction which asserts that the distinction of the fundamental was the last distinction!" (639). What the theory of the observation of observation holds, then, is that the world is not given—as in the traditional, representationalist frame—but is rather brought forth in the dynamic interaction of observer and observed. As Maturana and Varela put it, "Our intention is to bypass entirely this logical geography of inner versus outer by studying cognition not as recovery or projection but as embodied action" (*Tree* 172). But more than this, since that bringing forth takes place by means of paradoxical distinctions, it means, as Maturana and Varela put it, that "every world brought forth necessarily hides its origins. By existing, we generate cognitive 'blind spots' that can be cleared only through generating new blind spots in another domain. We do not see what we do not see, and what we do not see does not exist" (*Tree* 242).

But if the general paradigm of systems theory has migrated from the natural and mathematical sciences to exert considerable influence in the social sciences (as the work of Niklas Luhmann demonstrates), such has not been the case, for the most part, in the humanities, where scholars have been slow to recognize the potentially productive and liberating implications of the breakdown of the representationalist world view. There are, we believe, two primary reasons for this. First, most of the recent work on new developments in the sciences that has garnered much attention in the humanities and, more broadly, in cultural studies generally— such as the works by Bruno Latour and Steve Woolgar's *Laboratory Life*, Evelyn Fox Keller's *Secrets of Life, Secrets of Death*, or Donna Haraway's *Simians, Cyborgs, and Women*—have focused for the most part on the cultural, ideological, textual, and rhetorical dimensions of scientific knowledge, often leaving aside the potential for mutual illumination of what C. P. Snow called "the two cultures" on general epistemological questions of interpretation. And while a handful of admirable analyses by N. Katherine Hayles, Brian Porush, Brian Massumi, Paisley Livingston, Hans Ulrich Gumbrecht, George Levine, Michel Serres, and others have paid increasing attention (often under the growing rubric of "science and literature") to the cross-fertilization of literary and scientific theories, the more prevalent response of the humanities to these new scientific developments has been largely recuperative—that is, to reduce the problems raised by them to problems of textuality, rhetoric, theme, and style, all of which can then be analyzed by the tried-and-true methods of a more or less traditional literary criticism. The importance of emphasizing the textual and rhetorical construction of scientific knowledge should not be underestimated. Nevertheless, it has often tended to reinforce disciplinary boundaries rather than encourage vigorous interdisciplinary dialogue about the nature of knowledge and the problem of interpretation at the beginning of the twenty-first century.

This is unfortunate, of course, because the sciences and humanities at the current moment share a common set of fundamental epistemological problems, many of which we have already mentioned: logical circularity, paradoxical self-reference, unpredictable recursivity and self-organization, how to recognize contingency without lapsing into relativism, how to posit the essentially closed systematicity of the cultural or knowledge-making system and at the same time account for its historical change—just to name a few. And this, in turn, leads us to the second reason, mentioned above, why the humanities have been slow to assimilate the potentially productive implications of the new scientific paradigms for thinking through the breakdown of the representationalist world view. The epistemological challenges that the sciences have confronted by undertaking a fundamental paradigm shift have continued to be framed within the humanities by a rather disputatious and unproductive standoff between two poles of a reductive dualism: realism and idealism, objectivity and relativism, knowledge and ideology, and so on. As the philosopher Richard Rorty explains in his study *Objectivity, Relativism, and Truth,* the intellectual habits that have governed these discussions have remained grounded in representationalism's assumption that "'making true' and 'representing' are reciprocal relations," wherein the "item which makes S true is the one represented by S," and in realism's "idea that inquiry is a matter of finding out the nature of something which lies outside the web of beliefs and desires," in which "the object of inquiry—what lies outside the organism—has a context of its own, a context which is privileged by virtue of being the object's rather than the inquirer's" (4, 96). The problem with this framework, of course, is that it really gives us no way to rigorously articulate what most of us believe: that, as Rorty puts it, "it is no truer that 'atoms are what they are because we use "atom" as we do' than that 'we use "atom" as we do because atoms are as they are'" (5).

The limitations of the representationalist frame for dealing with the problems of knowledge raised by the new scientific paradigms are especially clear in the ongoing debates over the status of what is called "postmodernism." On the one side, as we have already noted, we find critics of diverse political stripe who lament that with the breakdown of the realist worldview, we experience what Michel Foucault calls the "death of the subject" and that the loss of meaning that undermines the philosophical, ethical, and political promises of the project of modernity and, beyond that, of the Enlightenment itself. On this view, in the absence of realism we can no longer make a case (to borrow E. D. Hirsch's well-worn phrase) for "the validity of interpretation," wherein the interpretive act depends for its veracity—and ultimately for its ethical or political efficacy—on its representational adequation, its faithful mirroring, if you will, of the objective meaning of the text or social phenomenon. And on this view, if objectivity or something very much like it is not possible, then we are automatically driven back upon a self-refuting relativism and even nihilism. As one recent anti-antifoundationalist puts it, "How does one rule out categorical theories in principle without getting categorical? How does one universalize about theory's inability to universalize?" (Fairlamb 57). This epistemological objection, in turn, underwrites the sorts of political and ethical charges made by Marxian critic Norman Geras, who still grapples with the dilemma Horkheimer and Adorno identified a half century ago: "If there is no truth, then there is no injustice. Stated less simplistically, if truth is wholly relativized or internalized to particular discourses or language games or social practices, there is no injustice. . . . Morally and politically, therefore, anything goes" (110).

On the other side, we find proponents of postmodernism who accept or even celebrate this very loss of representational veracity as a liberation of philosophical, social, and cultural analysis from what Jacques Derrida has famously called "logocentrism" and the disciplinary practices of knowledge legitimation and production that enforce it. As Malcolm Ashmore, Derek Edwards, and Jonathan Potter have argued recently, "If objective truth and validity are renounced in favor of social process and practical reasoning, then so also must be any notion of a commitment to '*equal* validity.' Far from ruling out the possibility of justification of a particular view, relativism insists upon it" (10). After all, they remind us, epistemological realism of the sort promoted by Geras "is no more secure than relativism in making sure the good guys win, nor even in defining who the good guys are—except according to some specific realist assumptions that place such issues outside of argument" (11). Meanwhile, most scholars have settled for an uneasy position somewhere in the middle of these two poles: that there is indeed a preexistent, finite reality with its own objective nature, but one that is viewed differently by different observers according to the cultural and social determinations that shape their particular view of things. The problem with this commonsensical view, however, is that it is purchased at the expense of incoherence, since it simultaneously endorses and disavows the very representationalism that it bridles against.

In our view, then, the challenge the humanities now face is not to encourage the taking of this or that position within this epistemological framework, but to rethink the representationalist framework itself. While this project is underway in some of the more prominent varieties of contemporary philosophical and social theory usually associated, vaguely, with "pragmatism" or "deconstruction," we feel that systems theory can offer a fresh perspective on the concerns addressed by these styles of thinking because it differs significantly from all of these. Pragmatism, like systems theory, insists upon the social and historical contingency of all knowledge; but unlike pragmatism, systems theory believes that the recognition of this contingency requires an ongoing commitment to questions of epistemology, rather than—to use Cornel West's characterization—the "evasion of epistemology-centered philosophy." And unlike deconstruction (which is suspicious of both universal claims and utilitarian thought in general), systems theory attempts to provide a rigorously coherent means of describing all systems, whether organic or inorganic, and quite unabashedly serves a wide range of practical applications in a variety of fields. Finally, systems theory—by providing analyses that are not differentiated along the lines of the traditional dichotomies governing thought that ground both idealism in its Kantian and Cartesian forms and materialism in its Marxist form (subject/object, human/nonhuman, culture/nature, organic/mechanical)—offers the possibility of a theory of knowledge that can account with greater range and power for the intrication of human beings in what Bruno Latour, in his recent study *We Have Never Been Modern,* calls the "hybrid networks" of social, informational, and ecological systems in which we will find ourselves increasingly enmeshed in the coming years.

Systems Theory and the Environment of (Post)Modernity

It is in helping us to understand and critically address Latour's "hybrid networks" of postmodernity that systems theory might most immediately— and, as it were, thematically—seem to be of political use, not least because systems theory makes use of the same formal and dynamic models *across* what have been viewed traditionally as discrete ontological domains (organic versus mechanical, natural versus cultural, and so on). The usefulness of systems theory's ability to "deontologize" and transgress such boundaries ought to be clear for "new social movements" such as environmentalism (the relationship between population genetics and endangered species protection, for instance, or how the ecology of wetlands preservation outstrips legal property boundaries), animal rights (the effects of taylorized and chemical-intensive "factory farming" on animals, the environment, family farms, and public health), feminism (the impact of the medical establishment and its technologies, especially reproductive technologies, on women), and gay rights (same-sex marriage in the legal system, for example, or how the ecology of AIDS is related to the spatial distribution of both gay and straight sexual practice). As we have already suggested, however, the left's response to systems theory has typically been a rather politically ambivalent one. On the one hand, critics such as Carolyn Merchant and Peter Galison find in systems theory "the apotheosis of behaviorism," making "an angel of control and a devil of disorder" (Galison 251, 266). As Merchant puts it, systems theory can "be appropriated, not as a source of cultural transformation, but as an instrument for technocratic management of society and nature, leaving the prevailing social and economic order unchanged" (104).

Some, however, have felt that systems theory might very well be politically and ethically enabling—and in any case they have reminded us that it is an inescapable, constitutive feature of our experience. As Donna Haraway puts it in her well-known "Manifesto for Cyborgs," it may be that "a cyborg world is about the final imposition of a grid of control on the planet," but

From another perspective, a cyborg world might be about lived social and bodily realities in which people are not afraid of their joint kinship with animals and machines, not afraid of permanently partial identities and contradictory standpoints. The political struggle is to see from both perspectives at once because each reveals both dominations and possibilities unimaginable from the other vantage point. (154)

These political uses and implications of systems theory will continue to be under dispute, of course, for the foreseeable future. What we wish to focus on here, however, is a somewhat different angle of vision on the relationship between systems theory and politics. For the point of this collection is not to offer suggestions about political solutions to the challenges posed by contemporary society, nor is it to reach some final decision about the good or bad political status of systems theory, still less is it to show how systems theory can provide a new and improved grounding for political practice of one sort or another. These are all very important and pressing issues, of course, and they are addressed in some detail in these pages. But they are, in the end, tertiary to the more fundamental engagement of politics that we see emerging out of this collection, an engagement that takes place on the terrain of the *conditions of possibility* for politics itself.

The concept of politics as it is usually mobilized in critical theory presumes a whole cluster of "truths"—about subjects lucidly and transparently recognizing their "interests" (or classes of people serving some privileged relation to historical process or public space, whether they recognize those interests or not); about the necessity of praxis being grounded in normative concepts underwritten by philosophical foundationalism; about the intersubjectivity upon which mass movements depend; about the perfectibility of the human condition—which are more or less explicitly under dispute in these pages, and are called into serious question, we would add, by developments associated with postmodernity itself. So our aim is not to provide political *answers,* but rather to use systems-informed thinking to reopen politics as a *question.*

If, therefore, we wish to investigate the conditions of possibility for politics in the (post)modern world, we must come to terms with the structure of modernity as the space in which the political can be found. We can agree with all the diagnosticians of modernity, from Kant to Habermas and Luhmann, that this space is fractured and marked by difference. We can agree, in other words, with Habermas when he characterizes the various projects of modernity as a set of responses to the "problem of modernity's *self-reassurance*," the problem of an "anxiety caused by the fact that a modernity without models had to stabilize itself on the basis of the very diremptions it had wrought" (16). And we can thus agree with conventional descriptions of modernity as marked by the "loss" (or simply the increasing impotence) of successive transcendental justifications (God as Reason and later, under Enlightenment, Reason as God) under the spur of increasing internal differentiation into separate, immanent spheres of knowledge, where none can lay absolute claim to authority over the operations or self-understanding of any other.

However, rather than search for a new form of universality that will somehow compensate for this perceived loss, we can accept Luhmann's dispassionate description of this differentiation of modern society without regret and without nostalgia, and we thus take seriously his rejection of the so-called "project of modernity" and its desire for a universally assumed but nowhere concretely localizable lifeworld (even if only formally or procedurally assumed) from which the differentiated language games and social systems of postmodernity can be normatively steered—a rejection that Luhmann, of course, shares with many of the more well-known theorists of postmodernity discussed in these pages. The inability to think the unity of reason in the face of the proliferation of self-referential language games and social systems, in other words, is simultaneously the inability to locate a space "outside the walls of the city" (Benhabib 28) from which society can be observed or recreated as a unified whole. If one views the differentiation of reason and modern society as fragmentation, reification, and alienation, and sees theories (such as Luhmann's) that "describe" rather than "critique" it as skeptical, relativistic, and nihilisitic—in short, if one views the modern condition as *tragic*—then politics must be seen as both the book and the sword that will redress our fallen state. It must theorize community, consensus, holism, and seek to establish the conditions for community's possibility by transcending—by means of theory, history, class, lifeworld, or any of a number of other devices—the limitations currently in place. If one views differentiation negatively, in other words, politics must compensate for the lack that differentiation represents. If modernity is ill, and health is the standard, then politics is the cure.

Such a view of politics, however, will not be the one taken here. On the contrary, the antinomy of theoretical and practical reason and the differentiation of society into competing value spheres and social systems will be taken as the modern condition of possibility for politics. Accordingly, we suggest that the trope of politics as *critique* be rethought in terms of politics as *conflict*. This is not to say that complacency is to replace concrete criticisms of social ills, but rather that the pathos of eschatology and the ethos of salvation that clings to calls for radical transformation of society give way to a dynamic structure of political contest in which God, History, and the appropriated Marginal Other are on no one's side. The question to be addressed, then, is this: Given modernity, how is politics possible? We consider politics to be an activity that takes place within the space of modernity, and not one that forever claims a place outside its walls in an attempt to bring those walls down.

Indeed, now that the dust has settled, the controversies surrounding the discussion of postmodernity have allowed us to look not *beyond,* but rather *at* modernity with fresh eyes, not as a quasi-teleological project, but as an ever-expanding zone of complexification and conflict. We may not wish to sing the praises of the "performativity of the system" (to use Lyotard's phrase from *The Postmodern Condition*), but we must also realize that even critique increases the system's performativity. In this light, what are typically viewed as modernity's fatal flaws—its systemic "reifications" and differentiations, its theoretical relativism, its "nihilistic" loss of tradition and skeptical questioning of the true, the good, and the beautiful, its "paralytic" antinomy between theory and practice, word and deed—may be seen instead as the enabling conditions of those activities, political and otherwise, that take place within a modernity whose most enabling fiction and abiding fantasy is that we can step outside or transcend it.

As Albrecht Wellmer, in the introduction to his pointedly titled volume *The Persistence of Modernity,* notes:

> If there is something new in postmodernism, it is not the radical critique of modernity, but the redirection of this critique. With postmodernism, ironically enough, it becomes obvious that the critique of the modern, inasmuch as it knows its own parameters, can only aim at expanding the interior space of modernity, not at surpassing it. For it is the very gesture of radical surpassing—romantic utopianism—that postmodernism has called into question. (vii)

Here, Wellmer presents us with the picture of a resilient and unreconciled modernity as "an unsurpassable horizon" (vii). All attempts to move beyond the modern landscape have proven futile—indeed, they have provided the energy for extending the domain or the "interior space" of that which was to be left behind. Postmodernism, then, rather than some supersession of modernity, "at its best might be seen as a self-critical—a sceptical, ironic, but nevertheless unrelenting—form of modernism; a modernism beyond utopianism, scientism, and foundationalism: in short, a post*metaphysical* modernism" (vii) that takes as its primary target not modernity itself, but modernity's ill-advised projects of self-transcendence. Thus, if the inability to overcome modernity is postmodernism's triumph, it is because it marks the end of some of modernity's most treasured illusions.

Our sense of the "political," then, has much in common with what, in recent years, has been variously labeled "agonistic liberalism" (John Gray), "agonistic pluralism" (Chantal Mouffe), or the "ethos of pluralization" recommended by William E. Connolly. As the political philosopher Ernesto Laclau argues,

> recognition of the historicity of being—and thus of the purely human and discursive nature of truth—opens new opportunities for a radical politics. Such opportunities stem from the new liberty gained in relation to the object and from an understanding of the socially constructed nature of any objectivity. (4)

For us, then, the political resonance of systems theory at the current moment might best be viewed in light of a more general impulse toward a radical political pluralism, what Steven Best and Douglas Kellner call a "multiperspectival social theory" that sees each interpretive act, each observation, as "a way of seeing, a vantage point or optic to analyze specific phenomena," via a type of theory that is "reflexive and self-critical, aware of its presuppositions, interests, and limitations," a theory "nondogmatic and open to disconfirmation and revision, eschewing the quest for certainty," and "open to new historical conditions, theoretical perspectives, and political applications" (264–65, 257–58).

It is here that the resolute commitment of systems theory to contingency and the partiality of all knowledge may be viewed as the hallmark of a genuinely pluralist philosophy, for it is only in the distributed, constitutively incomplete observations of different observers that a critical view of any observed system, or any social fact, can be constructed. In Luhmann, for instance, the insistence on the inescapable "blind spot" of observation—the fact that observation is always based upon a paradoxical distinction that it cannot critically disclose and at the same time carry out its operation—leads rather directly to our recognition of the essential aporia of any authority that derives from it (the authority, say, of the system that enforces the distinction legal/illegal). From this vantage, the model of second-order observation in systems theory may be fruitfully paralleled with the theory of democratic "social antagonism" in the work of Ernesto Laclau, Chantal Mouffe, and Slavoj Zizek. These theorists, like Luhmann, do not disavow or repress what Zizek calls the "broken and perverted" (or Luhmannian paradoxical and tautological) nature of communication, but rather derive from it the conditions of possibility for democratic sociality as such. As Zizek puts it, "the limitation proper to the symbolic field as such" is "the fact that the signifying field is always structured around a certain fundamental deadlock" (or what Luhmann characterizes as the "blockage" of paradoxical self-reference). Like the theorists of social antagonism, Luhmann insists that the distribution or unfolding of such "blockages" or paradoxes is not an impediment to democratic society but is in fact absolutely crucial to it. And hence, a truly pluralist, "multiperspectival" philosophy should avoid at all costs the quintessentially modernist and Enlightenment strategy (of the sort we find in Habermas) of reducing complexity via social consensus.

Moreover, just as systems theory holds that the environment is always already more complex than any and all systems and the observations and operations they carry out—and hence serves as a permanently destabilizing and dynamic field of perturbations to which the system must constantly adapt through reductive coding—so the theory of social antagonism holds that

> antagonism is the limit of all objectivity. This should be understood in its most literal sense: as the assertion that antagonism does not have an objective meaning, but is that which prevents the constitution of objectivity itself. The Hegelian conception of contradiction subsumed within it both social antagonisms and the processes of natural change. This was possible insofar as contradiction was conceived as an internal moment of the concept; the rationality of the real was the rationality of the system, with any "outside" excluded by definition. In our conception of antagonism, on the other hand, we are faced with a "constitutive outside." It is an "outside" which blocks the identity of the "inside" (and is, nonetheless, the prerequisite for its constitution at the same time). With antagonism, denial does not originate from the "inside" of identity itself but, in its most radical sense, *from outside*. (Laclau 17)

If this is so, Laclau argues, "if the constitutive nature of antagonism is taken for granted, the mode of questioning of the social is completely modified, since contingency radically penetrates the very identity of the social agents. The two antagonistic forces are not the expression of a deeper objective movement that would include both of them; and the course of history cannot be explained in terms of the essential 'objectivity' of either. The latter is always threatened by a constitutive outside" (22).

From this vantage, systems theory, because of its epistemological rigor, its avowed posthumanism, and its resolute antifoundationalism, may be seen as an especially valuable partner in the project of theorizing the conditions of possibility for a postmodernist political pluralism, precisely by helping us to realize that theory cannot ground politics in the way that modernity imagines because—to borrow Rodolphe Gasché's formulation—the conditions of possibility identified by theory are "at the same time conditions of impossibility." What Gasché calls the "infrastructures" of theory—he has in mind the Derridean notions of "trace," "*différance*," and so on—enable the possibility of the Foucauldian "permanent critique" associated with modernity precisely because they *disable* it (4) by failing to ground or secure it. As Gasché explains—and here the similarities with systems theory are striking indeed—"The law articulated by an infrastructure applies to itself as well. It has an identity, that is, a minimal ideality that can be repeated only at the price of a relentless deferral of itself." (7). Like the law of the paradoxical identity of any constitutive distinction, which Luhmann borrows from George Spencer Brown, "What these laws establish, indeed, is that any ideality, identity, or generality hinges on a prior doubling, pointing away from (self), and referral to an Other—in other words, on a prior singularization" (7). This does not mean, however, that the affirmation of difference pure and simple is enough, for as Gasché explains in his discussion of Heidegger's concept of *Versammlung* (or "gathering"), "To reject all gathering because it can turn into self-identical individuality, totality, or System is to close the doors of reflection and philosophical interpretation. Is this not to abort what gathering still holds out for the future, to reveal a lack of respect for what is to come, for what has never yet been present?" (20).

It is on the basis of this reorientation of theory toward its conditions of (im)possibility, then, that theory can make good on the critical imperatives described in Foucault's reassessment of Enlightenment, because it is on this basis that the relationship between theory and a future yet to come can be reoriented away from dialectical closure—away from, as Derrida famously puts it in "Structure, Sign, and Play," *arche* and *telos*—and toward what Luhmann characterizes as the operationalization of difference, which the dialectic *says* it values but *can* value—as Adorno himself realized—only in its endless deferral of difference (hence the necessity, for Adorno, of *negative* dialectics). In this light, systems theory may be seen, as Luhmann puts it, as something like the "reconstruction of deconstruction" ("Deconstruction as Second-Order Observing" 770) insofar as it examines the pragmatic effects of reorienting theory away from dialectic and toward the difference of identity and nonidentity, and shows how the failure of identity-based schemes may be operationalized systemically as a kind of necessity and success, and not only by systems that are either language- or text-based—that is to say, not only by systems that are human and/or human*ist*.

The priority of systems theory in the "reconstruction of deconstruction" is suggested in a second, different register by Pierre Bourdieu's passing criticism of deconstruction in *The Field of Cultural Production*: that "by claiming a radical break with the ambition of uncovering ahistorical and ontologically founded essences, this critique is likely to discourage the search for the foundation" of social forms and institutions where they are "truly located, namely, in the history" of those forms and institutions (255). Luhmann has provided his own version of this historical emergence, of course, in his theory of "functionally differentiated" society. According to Luhmann, the transition to modern society is characterized above all by the movement away from stratified or hierarchical organization, in which the absolute monarch, the church, the court, or the aristocracy represents society as a whole, and toward a society of operationally closed, self-referential function systems—the legal system, the economic system, the education system, and so on. Under functional differentiation, as Dietrich Schwanitz puts it, "society is no longer regarded as the sum of its parts, but as a combination of system-environment differentiations, each of which reconstructs the overall system as a unity for the respective subsystem and its specific environment according to the internal boundary of the subsystem" (144). Luhmann's account of functional differentiation gives us a picture of modern (and postmodern) society as a horizontal plane on which the different autopoietic function systems exist side by side, with no one system (the economic in the Marxist account, say) able to overdetermine the others. "The present state of world society," Luhmann writes in "Why Does Society Describe Itself as Postmodern?" "can hardly be explained as a consequence of stratification. The dominant type of system-building within contemporary society relates to functions, not to social status, rank, and hierarchial order. The so-called 'class society' was already a consequence of functional differentiation, resulting, in particular, from the differentiation of the economic and the educational subsystems of society."

Luhmann's account of our current situation recalls much that readers will have already seen in Fredric Jameson's groundbreaking discussion of postmodernism—not least of all the collapse of "depth models" of knowledge and the loss of "critical distance" whereby one could assume a privileged perspective on difference and hypercomplexity. But if Jameson's Hegelian dialectic, epistemological privileging of Marxist "science," and explanatory priority of the economic as determinative is, for us, untenable, his model of the postmodern nevertheless sharply reveals at least one problematic aspect of Luhmann's account—namely, the role of the economic system in modern society. In dealing with the influence of economics on politics, pedagogy, culture, and a whole host of additional social realms, it makes a world of difference, of course, what concept of society and the social one deploys. In the world of Jameson's "late capitalism," the economic system remains at the center of social organization, and hence all explanation of that organization must start with the economic. From such a perspective, Luhmann's contention that the functionally differentiated systems of society are horizontally ordered sounds hardly plausible. Are not the social problems with which we are all familiar related to the overdetermining fact of a dramatic redistribution of wealth upward from the working and middle classes to the wealthy since the late 1970s? Luhmann's relative silence regarding the force that the economic system exerts on the political system (for instance, in the form of political action committees and campaign finance) or on the legal system would seem only to confirm a central weakness in his theory. But if, on the other hand, one starts with the assumption of a nonhierarchical order of functional differentiation, then the centripetal force of the economy becomes not the symbol of the established order, but rather a sign of the threat of de-differentiation—in which case the political battle is not one of replacing one sort of hierarchically overdetermining economic system with another (capitalist with communist, say), but rather limiting and constraining the economic as such so as to preserve the structure (namely differentiated complexity) of society itself. On this view, such "preservation" of the structure of difference is only possible by replicating difference, dissent, and conflict wherever "hegemony" (to use a classical concept) threatens—whether it is hegemony of the economic, the moral, or any other discourse or system.

    As Jameson would no doubt be the first to point out, Luhmann's account reproduces all the problems of a liberal technocratic functionalism that has no way to address the sharp asymmetries of power in the social field, asymmetries that make the autopoiesis of social systems work better for some than for others. To say that social differentiation is perforce good is immediately to beg the question raised by Best and Kellner:

that "some people and groups are in far better positions—politically, economically, and psychologically—to speak than others" (288). For Luhmann, of course, such exclusions and silences are a matter of course, because no social whole in which exclusion can be excluded is functionally feasible (nor does modernity provide any examples of such a society). Hence, for Luhmann, to hold out a vision of such a whole is an illusion— an "ideology," one is tempted to say—of the most misleading sort. Nevertheless, what is pressed by Jameson's analysis is once again the issue of structural causality (as logically problematic or theoretically challenging as it may be), which wagers that some social systems (in this case, the economic) exert—at least for the time being—more overdetermining force on the social field as a whole than others. And what this means is that systems theory will not be able to maximize its own political utility until it confronts inherent inequities of power that complicate and compromise the *formal* equivalence of different observers in the social field. We need to remember, as Jameson puts it, that "no matter how desirable this postmodern philosophical free play may be, it cannot now be practiced; however conceivable and imaginable it may have become as a philosophical aesthetic (but it would be important to ask what the historical preconditions for the very conception of this ideal and the possibility of imagining it are), anti-systematic writing today is condemned to remain within the 'system'" of global capitalism and the law of value (*Late Marxism* 27).

For Jameson—though also not exclusively for Marxists—we do not in fact live in a fully functionally differentiated society organized horizontally, but instead in a kind of hybrid society in which highly autonomous and self-referential forms usually associated with postmodernism (especially in areas like media and communications) coexist alongside more traditional hierarchical ones (such as the economic and class systems) that are associated with the stratified societies of early modernity, and that exercise an asymmetrical influence on the autopoiesis of other social systems. In such an account, the relation between the modern and the postmodern—between the twin Enlightenment legacies of administered society and permanent critique, Jameson's "scientific" totalization and Luhmann's liberal utopian vision of full functional differentiation, Gasché's conditions of possibility and impossibility—might therefore be redescribed in terms (following Raymond Williams) of dominant, residual, and emergent historical trends (121–27). On this view, as Best and Kellner put it, "we might want to speak of postmodern phenomena as only emergent tendencies within a still dominant modernity"—a modernity, to be sure, that follows Jameson's hierarchial account rather than Luhmann's differentiated one—"that is haunted as well by various forms of residual, traditional culture, or which intensify key dynamics of modernity, such as innovation and fragmentation" (279). Such a hybrid account would present a more persuasive and compelling picture of our current situation than either Jameson's *Marxist utopian* totalization in the name of the economic or Luhmann's *liberal utopian* view of functional differentiation taken singly.

But from what vantage can one engage in a historicization of the emergent differences we find between a Jamesonian critique and a Luhmannian one? It is here that Jameson's famous dictum in *The Political Unconscious*— "always historicize!"—has, if rigorously pursued, unexpected consequences for his own account, and for the dialectical model upon which it is based. This is so, as Lyotard argues in some detail in *The Postmodern Condition,* because of changes in the conditions of knowledge themselves under postmodernism—changes which, as Barry Smart has correctly pointed out, the Marxist tradition has by and large ignored or treated with insufficient attention. As Smart argues, chief among these are what Anthony Giddens calls—in a formulation reminiscent of Luhmann's "hypercomplexity"—the "reflexivity or circularity of social knowledge" (Smart 193). In fields such as cognitive science, for example, it is not simply that changes in the social conditions of knowledge—in technologies, practices, and the very material factors of knowledge production in which Marxism should be interested—change how knowledge procedures are conducted; it is rather that they in turn transform what knowledge is and how we may interact with and use it.

It is perfectly possible, indeed it is entirely necessary, to make the historical materialist point that the paradigms of cybernetics arise out of the specific social, economic, and political conditions of the World War II effort, and more specifically in logistics and weapons research—that they are, in a sense, only possible in such a context (leaving aside the more daunting Marxist problem of whether these developments can be shown to be determined by the economic in the last instance!). But the larger point is that even if we proceed along the lines of a more or less traditional form of materialist historicization, the changes we will identify in the social conditions and production of knowledge do not leave the quality and character of that knowledge untouched. In the case of complexity theory, for example, new paradigms of knowledge—ones with demonstrable pragmatic power and value, as the use of chaos theory in cardiology, meteorology, and economic analysis makes clear—do not

simply question but fundamentally undermine the subject/object paradigm upon which dialectics and the dialectical account of causality depends. How can we continue to believe in anything like a Marxist "science" when the very foundations upon which that science bases itself
have been radically questioned if not rendered obsolete by changes in the
social conditions of knowledge and the new theoretical developments—
like chaos theory, complexity theory, and systems theory—they have
made possible? Nor are these changes limited to epistemological and
philosophical consequences, for as Giddens points out, what they mean
is that "[n]o matter how well a system is designed and no matter how
efficient its operators, the consequences of its introduction and functioning . . . cannot be wholly predicted. . . . New knowledge (concepts, theories, findings) does not simply render the social world more transparent,
but alters its nature, spinning off in novel directions. . . . For all these reasons, we cannot seize 'history' and bend it readily to our collective purposes" (153–54).

The material and historical conditions highlighted by Giddens, and
the challenges they pose for "modeling praxis without subjects and objects," is refocused in light of the relationship between Romanticism and
modernity in Marjorie Levinson's contribution to this volume. Levinson's
political deployment and rethinking of key concepts from Hegelian and
Marxist thought is flanked by two exchanges. The first, between Niklas
Luhmann and Peter Uwe Hohendahl, examines the relationship between
modernity and postmodernity—and the relative successes and failures
thereof—in view of key conceptual linchpins of systems theory, such as
second-order observation and functional differentiation. The second,
between William Rasch and Drucilla Cornell, interrogates the very idea
of ethics by contrasting the views of Luhmannian systems theory and
Derridean deconstruction on the relationship between the idea of justice
and the legal system.

Section two of this collection provides a critical and refreshingly accessible overview of some of the major concepts and thinkers of systems
theory—and where better to begin than with a conversation with one of
its major practitioners, Niklas Luhmann, and one of the leading figures
in science and literature studies, N. Katherine Hayles? Many of the problems and concepts raised in the interview find more fully articulated and
detailed treatment in the essays of Hayles and Wolfe, which provide
somewhat contrasting views (but also rather similar political and ethical
reservations) about the enormously important work in systems theory and
epistemology of Chilean biologists Humberto Maturana and Francisco

Varela. The "crash course" in systems theory initiated by the first two sections of this collection takes readers to rather more familiar terrain in the third and final section, where some of the key theoretical, methodological, and historical commitments of systems theory are brought into explicit parallel with some of the major theorists associated with postmodernism. While Rasch compares the rather different posture toward the relationship between modernity and exclusion in Luhmann and Jean-François Lyotard, Jonathan Elmer turns to the discussion of modernity in Michel Foucault, and even more to Jacques Lacan's conjugation of psychoanalysis and cybernetics, for a critical vantage on Luhmann's discussion of the problem of communication and the relationship between psychic and social systems. Wolfe, meanwhile, moves not to psychoanalysis but to pragmatism—specifically the pragmatism of Richard Rorty—in an examination of how systems theory may be used to renovate pragmatism and its shared commitment to antifoundationalist theory, even as systems theory has, so far, remained politically landlocked within the same complacent liberalism that so many have noted in Rorty. After such a resolutely antiontological pairing as Rorty and Luhmann, it is only fitting that we conclude with a strong return to ontology—albeit of a distinctly postmodern sort—in Brian Massumi's deployment of the work of Gilles Deleuze to rethink concepts familiar to us from systems theory, such as emergence, virtuality, and event, concepts that may help us understand how perception, affect, and the bodily life more generally are increasingly central, increasingly hardwired if you will, to the politics of postmodernity.

Works Cited

Ashmore, Malcolm, Derek Edwards, and Jonathan Potter. "The Bottom Line: The Rhetoric of Reality Demonstrations." *Configurations* 2, no. 1 (winter 1994): 11–34.

Bateson, Gregory. *Steps to an Ecology of Mind: A Revolutionary Approach to Man's Understanding of Himself.* New York: Ballantine, 1972.

Benhabib, Seyla. "Feminism and Postmodernism: An Uneasy Alliance." In *Feminist Contentions: A Philosophical Exchange,* by Seyla Benhabib, Judith Butler, Drucilla Cornell, and Nancy Fraser. New York: Routledge, 1995.

Best, Steven, and Douglas Kellner. *Postmodern Theory: Critical Interrogations.* New York: Guilford, 1991.

Bourdieu, Pierre. *The Field of Cultural Production.* Edited by Randal Johnson. New York: Columbia University Press, 1993.

Connolly, William E. *The Ethos of Pluralization*. Minneapolis: University of Minnesota Press, 1995.

Derrida, Jacques. *Writing and Difference*. Translated by Alan Bass. Chicago: University of Chicago Press, 1978.

Fairlamb, Horace. *Critical Foundation: Postmodernity and the Question of Foundations*. New York: Cambridge University Press, 1994.

Galison, Peter. "The Ontology of the Enemy: Norbert Wiener and the Cybernetic Vision." *Critical Inquiry* 21, no. 1 (autumn 1994): 228–66.

Gasché, Rodolphe. *Inventions of Difference: On Jacques Derrida*. Cambridge: Harvard University Press, 1994.

Geras, Norman. "Language, Truth and Justice." *New Left Review* 209 (January/February 1995): 110–135.

Giddens, Anthony. *The Consequences of Modernity*. Stanford: Stanford University Press, 1990.

Glanville, Ranulph, and Francisco Varela. "Your Inside Is Out and Your Outside Is In (Beatles [1968])." In *Applied Systems and Cybernetics*, Proceedings of the International Congress on Applied Systems Research and Cybernetics. Vol. 2, *Systems, Concepts, Models, and Methodology*, edited by G. E. Lasker. New York: Pergamon, 1980.

Gray, John. *Isaiah Berlin*. Princeton: Princeton University Press, 1996.

Habermas, Jürgen. *The Philosophical Discourse of Modernity: Twelve Lectures*. Translated by Frederick G. Lawrence. Cambridge: MIT Press, 1987.

Haraway, Donna. *Simians, Cyborgs, and Women: The Reinvention of Nature*. New York: Routledge, 1991.

Heims, Steve Joshua. *Constructing a Social Science for Postwar America: The Cybernetics Group, 1946–1953*. Cambridge: MIT Press, 1993.

Hirsch, E. D. *Validity in Interpretation*. New Haven: Yale University Press, 1967.

Horkheimer, Max. *Between Philosophy and Social Science: Selected Early Writings*. Translated by G. Frederick Hunter, Matthew S. Kramer, and John Torpey. Cambridge: MIT Press, 1995.

Horkheimer, Max, and Theodor W. Adorno. *Dialectic of Enlightenment*. Translated by John Cumming. New York: Seabury, 1972.

Jameson, Fredric. *Late Marxism: Adorno, or, the Persistence of the Dialectic*. London: Verso, 1990.

———. *The Political Unconscious: Narrative as a Socially Symbolic Act*. Ithaca, N.Y.: Cornell University Press, 1981.

Keller, Evelyn Fox. *Secrets of Life, Secrets of Death: Essays on Language, Gender, and Science*. New York: Routledge, 1992.

Kuhn, Thomas S. *The Structure of Scientific Revolutions*. 2d ed. Chicago: University of Chicago Press, 1970.

Laclau, Ernesto. *New Reflections on the Revolution of Our Time.* London: Verso, 1990.

Latour, Bruno. *We Have Never Been Modern.* Translated by Catherine Porter. Cambridge: Harvard University Press, 1993.

Latour, Bruno, and Steve Woolgar. *Laboratory Life: The Social Construction of Scientific Facts.* Beverly Hills, Calif.: Sage, 1979.

Lilienfeld, Robert. *The Rise of Systems Theory: An Ideological Analysis.* New York: John Wiley and Sons, 1978.

Luhmann, Niklas. "Deconstruction as Second-Order Observing." *New Literary History* 24 (1993): 763–82.

———. "The Cognitive Program of Constructivism and a Reality That Remains Unknown." In *Selforganization: Portrait of a Scientific Revolution,* edited by Wolfgang Krohn, Gunter Kuppers, and Helga Nowotny. Dodrecht: Kluwer, 1990.

Lyotard, Jean-François. *The Postmodern Condition: A Report on Knowledge.* Translated by Geoff Bennington and Brian Massumi. Minneapolis: University of Minnesota Press, 1984.

———. *The Differend: Phrases in Dispute.* Translated by Georges Van Den Abbeele. Minneapolis: University of Minnesota Press, 1988.

Maturana, Humberto R., and Francisco J. Varela. *The Tree of Knowledge: The Biological Roots of Human Understanding.* Rev. ed. Boston: Shambhala, 1992.

Merchant, Carolyn. *Radical Ecology.* New York: Routledge, 1993.

Mouffe, Chantal. *The Return of the Political.* London: Verso, 1993.

Rorty, Richard. *Essays on Heidegger and Others.* Cambridge: Cambridge University Press, 1991.

———. *Objectivity, Relativism, and Truth.* Cambridge: Cambridge University Press, 1991.

Schwanitz, Dietrich. "Systems Theory According to Niklas Luhmann—Its Environment and Conceptual Strategies." *Cultural Critique* 30 (spring 1995): 137–70.

Smart, Barry. *Modern Conditions, Postmodern Controversies.* London: Routledge, 1992.

Spencer Brown, George. *The Laws of Form.* 2d ed. New York: Dutton, 1971.

Strauss, Leo. *What Is Political Philosophy?* Chicago: University of Chicago Press, 1988.

von Foerster, Heinz. *Observing Systems.* 2d ed. Seaside, Calif.: Intersystems, 1985.

Weaver, Warren. "Science and Complexity." *American Scientist* 36 (1948): 536–44.

Wellmer, Albrecht. *The Persistence of Modernity: Essays on Aesthetics, Ethics, and Postmodernism.* Cambridge: MIT Press, 1991.

West, Cornel. *The American Evasion of Philosophy: A Genealogy of Pragmatism.* Madison: University of Wisconsin Press, 1989.

Wiener, Norbert. *Cybernetics; or Control and Communication in the Animal and the Machine.* 2d ed. Cambridge: MIT Press, 1948. Reprint, 1961.

Williams, Raymond. *Marxism and Literature.* Oxford: Oxford University Press, 1977.

Zizek, Slavoj. "Beyond Discourse-Analysis." In *New Reflections on the Revolution of Our Time,* by Ernesto Laclau. London: Verso, 1990.

I

SYSTEMS THEORY AND/OR POSTMODERNISM?

HISTORICAL, POLITICAL, AND ETHICAL FRAMES

1. Why Does Society Describe Itself as Postmodern?

Niklas Luhmann

I

The discussion about modern or postmodern society operates on the semantic level. In it, we find many references to itself, many descriptions of descriptions, but hardly any attempt to take realities into account on the operational and structural level of social communications. Were we to care for realities, we would not see any sharp break between a modern and a postmodern society. For centuries we have had a monetary economy and we still have it. Perhaps there are signs that indicate a new centrality of financial markets, of banks and of portfolio strategies, that marginalize money spent for investment and consumption. We certainly can observe worldwide dissolution of the family economies of the past in agriculture and handicraft production. But it is and remains an economic system differentiated by transactions that use money. We have also had, for centuries now, a state-oriented political system, and we still have it. We face undeniable difficulties in establishing a state everywhere as a local address for political communications, but there is no alternative visible. We have positivistic legal systems, unified by constitutions. There are in many countries many doubts about whether the law will be applied. We find many cases in which the distinction between legally right and legally wrong is disregarded and does not matter at all. But there is no other type of law in view. We do scientific research as before, although now more conscious of risks or other unpleasant consequences. And we send, wherever possible, our children to schools, using up the best years of their lives to prepare them for an unknown future. Our whole life depends on technologies, today more than ever, and again, we see more problems, but no clear break with the past, no transition from a modern to a postmodern society.

Hence, the first question may be: Why do we indulge in a semantic discussion that does not burden itself with realities?

My answer to this question (there may be others) requires some knowledge about complex self-referential systems. Such systems make and continue to make a difference between the system and its environment. Every single operation that contributes to the self-reproduction of the system—that is, in the case of society, every single communication—reproduces this difference. In this sense, societies are operationally closed systems. They cannot operate outside their own boundaries.

Nevertheless, the system can use its own operations to distinguish itself from its environment. It can communicate about itself (about communication) and/or about its environment. It can distinguish between self-reference and hetero-reference, but this has to be done by an internal operation.

Operational closure is a necessary condition for observations, descriptions, and cognitions, because observing requires making a distinction and indicating one side of the distinction and not the other. The other side, the unmarked side, can be anything that is, for the time being, of no concern. Such distinctions have to be made by the system within the system. For we cannot suppose an environment (or a world) where everything is multiplied by anything, a world where every observable item includes the exclusion of everything else, or a world in which every thing has the properties of the absolute spirit in Hegel's sense.

Paying attention to this condition of the capacity of observing, we can see that the system *makes* the *difference* between system and environment and *copies* that difference in the system to be able to use it as a *distinction*. This operation of reinventing the difference as a distinction can be conceived as a reentry of a form into the form, or the distinction into the distinguished (Spencer Brown 56f., 69ff.).

Such a reentry has remarkable consequences. The form of a reentry is a paradoxical form, because the reentering form is the same and is not the same. To describe reentries we need a distinction, but the distinguished is the same. In mathematical terms it is an equation, and equation means something like "to be confused with" (Spencer Brown 69). For real systems making the difference and observing it by distinguishing self-reference and hetero-reference the reentry appears as ambivalence. So, psychiatrists say about themselves and their clients: "We can never be quite clear whether we are referring to the world as it *is* or to the world as we *see* it" (Ruesch and Bateson 238). There are self-correcting mechanisms available, but these always presuppose a "reality" with an ambiguous status.

The question whether it is the world as it is or the world as observed by the system remains for the system itself undecidable. Reality, then, may be an illusion, but the illusion itself is real.

Now, what is true for the environment is also true for the other side of the distinction, i.e., for the observing system itself. The reentry produces an "unresolvable indeterminacy" (Spencer Brown 57) of the system for itself. The system remains intransparent to itself. It can observe and describe itself and it can switch from one observation to another and can use many incompatible self-descriptions (Löfgren). Hence, such self-referential systems are hypercomplex systems because they may use a variety of very different distinctions to indicate the unity of their complexity.

These results become even more irritating when we remember that the system is an operationally closed system and therefore its own product. Since its operations depend on structures that are themselves the product of its operations we can, following Maturana, describe such systems as structurally determined systems. The state of the system is always the result of its own operations. Cyberneticians would say that the system uses its own output as input. This, however, means that it becomes too complex to calculate itself. It creates in itself an enormous amount of combinatorial possibilities. It operates for simple mathematical reasons as a "non-trivial machine" in Heinz von Foerster's sense (see "Principles of Self-Organization" and "Wie rekursiv"). And the remarkable insight is that it becomes intransparent, incalculable, unreliable and at the same time resilient, *because it produces itself and, thereby, determines its own state*. It cannot know, it cannot compute itself—not because its states depend upon events in its environment but because it arranges for self-created uncertainty.

To cope with these consequences of a reentry of the internal/external difference in itself, the system needs and constructs time. It needs a *memory function* to discriminate forgetting and remembering. Its past is given as a highly selected present and, in this sense, as reality. And it needs an *oscillator function* to be able to switch from marked to unmarked states in all kinds of distinctions, in particular to switch from hetero-reference to self-reference and vice versa. The system will not have an unselected past, nor will it be able to follow a linear prospect into the future. Its future will never become present, it cannot be marked by true statements. The relevant distinction, therefore, will not be true/false but something like flip/flop.

All these considerations apply to the societal system. The system is a nontrivial machine. It is an autopoietic system that produces and reproduces itself. It is a historical machine that has to start all its operations from a self-produced present state. It cannot calculate itself, but it can recursively connect memory and oscillation. It constructs distinguishable identities to reimpregnate its memory and to limit the range of possible futures. But it operates always in the present, never in its past and never in its future and always in the system and never in its environment.

This theory explains that we have to distinguish an operational level and a semantic level. The system is completely unable to calculate its operations in view of some representation of its own unity, or its end, or its complexity. But it *can* distinguish itself and describe itself, using a few of its operations to produce self-descriptions. For instance, it can say "we." It can refer to itself by a name. And it can use all kinds of complexity descriptions, e.g., differentiation. The self-designation of "modern" or "postmodern" society belongs to this category, and we understand now that this cannot be a representation, not even a map of the territory of its ongoing operations. It is just a way to organize or disorganize expectations.

But then, what can it possibly mean when Habermas says that society needs a "reasonable collective identity"?[1] An identity—distinguished from what?

For Habermas the answer seems to be clear—and simple: to distinguish itself from itself. This requires a normative concept of identity. Society is supposed to project a normative concept of rationality in order to compare itself with itself. The reasons for maintaining this split identity are—in spite of Habermas's attempt to relate them to linguistic authorities—historical, as was also the case in the famous Viennese lectures of Husserl. European Mankind or, in Habermas's case, the eighteenth century, invented the idea of self-critical reflection, and we are not supposed to drop this idea only because times are becoming rough and difficult. "Ideen sind Stärker als alle empirischen Mächte" ("Ideas are stronger than empirical powers"), in Husserl's words (335).

The norm of reasonable consensus or reasonable collective identity may be projected as unconditionally valid. It remains, however, distinguished from, and therefore conditioned by, what it rejects as an unsatisfactory state of present society, by what it characterizes as "crisis." Like all identities it is a double-sided form—it indicates the preferred state and thereby presupposes an undesirable state. The motive for choosing this and no other type of distinction is clearly stated: "But if modern societies have no possibility whatsoever of shaping a rational identity, then we are without any point of reference for a critique of modernity" (Habermas,

Discourse, 374). This claim, of course, is not true. We may critique modern society with regard to its probable consequences, its ecological consequences, its individual dissatisfactions, and still not need a "reasonable collective identity" to see the point. But Habermas does touch on the problematic identity of his guiding distinction—i.e., the identity of the difference between the norm and the deviant state of the system. This identity is the self-assured will to critique. Confronted with the necessity to found his own descriptions on the identity of his preferred distinction, he is forced to make a Gödelian jump and to make himself appear and disappear as an external observer.

II

We can neglect the theoretical differences between Husserl and Habermas, between the transcendental and the linguistic argument. The form of the projected identity is the same. It is a normative distinction and it can have only a historical and not a transcendental or a linguistic justification.[2] The identity of society distinguishes itself by pretending to be a norm. Society, then, is supposed to have a divided self—one acceptable and the other unacceptable.

The concept of "postmodernity" proposes a different solution for the same problem. It rejects any binding force of history—be it the European idea of self-critical philosophy, or the liberation (emancipation) of the individual as conceived in the eighteenth century. But then, what is this "postmodern" identity and what is excluded from it by distinction?

The term "postmodern" can accept many possible meanings. In one sense, introduced by Lyotard, the postmodern condition means the lack of a central unifying symbolization of the societal system, that is, the impossibility of a *métarécit* describing the unity of the system. In systems terminology this is nothing but "hypercomplexity," that is, the availability in the system of a plurality of descriptions of the system (Löfgren). This is, of course, neither new nor surprising. Ever since the French Revolution we have had this condition in Europe. Societal descriptions could focus on liberty or on equality, on institution or on organization, with good arguments for both sides; and for this very reason Max Weber refused to propose a concept of society.

Another meaning of postmodernity signifies the loss of the binding force of tradition. This, too, is so old that it has itself become a tradition (Winograd and Flores). The seventeenth century already rejected the idea that the validity of the law could be derived from a founding act, whether it be the justified Norman conquest for the common law (see

Hale), or a statute of the Emperor Lothar introducing the Roman civil law and canceling all other laws in the Holy Roman Empire.[3] The "origin" was going to be replaced by history itself. Indicating the origin *(arché, origo, Ursprung)* had been an easy answer to the "What is . . . ?" question, i.e., What is Being? What is a nobleman? What is the law? But if this was its function, it cannot be replaced with history. History grows older and older. It disappears in its past. It consumes itself. It accelerates to such a degree that there is not even time to ask the question "What is history?" and to look for an answer. History may have determined the present state of the system, but the result is typically dissatisfaction, need for revolution (either backward or forward) or at least reform, and in any case, a preference for difference.

How long, then, will Husserl and Habermas be able to maintain their old or modern idea of critical reasons without becoming conservatives who stick to a tradition that cannot maintain its identity but fades away? Has Plato ever been in Sicily? Has Habermas ever been in Bonn?

Postmodernity can also mean preference for inconsistency, that is, the praise of folly. But Erasmus remarked at the end of his *moriae encomium* that the praise of folly is itself foolish. It includes, as we would say, a performative contradiction, and Erasmus's conclusion is: An audience should be able to forget.

It may be good advice to forget postmodernity—but not before knowing what it has been. What, when, is the identity of the postmodern condition, which is to say: What is its specific difference?

Obviously, it has nothing to do with the structural drift or the evolution of modern society. Moreover, its description remains ambiguous. Perhaps it is an autological description, that is, a description that applies to itself. The description of postmodern society is itself a postmodern description, a description that includes its own performative "speech activity." If this is meant by "postmodernity," the term cannot say what it means, because this would lead to a confusion of constative and performative components of communication and in consequence—i.e., to its deconstruction.

Hence, we are again in a situation in which we have to cross the boundary of the form and to look at the other side. What is (or was) modernity so that postmodernity can be something else?

III

A sociological description of modern society will not start from the "project modernity," nor from the "postmodern condition." These are self-descriptions of our object, more or less convincing, two among many others (such as capitalist society, risk society, information society). Our object includes its own self-descriptions (including this one); for observations and descriptions exist only within the recursive context of communication that *is* and *reproduces* the societal system. But sociology can talk with its own voice.

The distinguishing (again: distinguishing!) characteristic of a sociological contribution to a self-description of society seems to be that it cannot neglect the operational and the structural level of societal reproduction. In other, more familiar words, sociology has to be an empirical science. The classical "sociology of knowledge" asked: What are the *relations* between the structural characteristics of a society (be it "capitalistic" stratification or division of labor in Durkheim's sense) on the one hand, and the forms of its knowledge on the other? Then, the truth value of statements relating to these relations could no longer be integrated with first-level knowledge, on which society bases its own communications. We therefore replace this relational phrasing by the distinction between operations and observations. Identity constructions meant to organize observations are always semantical artifacts. In hypercomplex systems they tend to become phantom identities. The interesting question then becomes: To be distinguished from what? And the answer will be: To be distinguished from difference and, in particular, from internal differentiation, produced and structured by the operations of the system.

To elaborate on this point I have to distinguish different forms of differentiation, namely stratification and functional differentiation. The history of societal self-descriptions shows very different constructions depending on whether the main form of differentiation is taken to be stratification or functional differentiation. But in both cases, the semantic artifact of system identity gets into trouble at the end of this century. Increasingly it becomes difficult to accept any description of identity when we have at the same time to accept the reality of differentiation and its consequences. *And this may be the reason why the idea of "postmodernity" attracts applause.*

The reaction to stratification began already in the seventeenth century when the order of estates lost its assumed foundation in nature and became an establishment created by the state and by law. Then, only the individual remained a natural entity and social order was thought to be an outcome of a contract, be it an undated one, a tacit one, or simply something that has to be assumed.[4] For more than a century the individual was thought to have an inborn capacity to be happy, regardless of his social status. Happiness for all became the remedy for the unjust distribution of wealth and power, but the condition was that the individual accepts his social position and does not aspire for more, as Molière's "bourgeois gentilhomme" did.

That society could not solve its problems as a community of happy individuals became evident during the second half of the eighteenth century, partly as a result of industrialization, but more as the consequence of the inclusion of agriculture into the monetary economy, the devastation of Scotland, the new poverty, and, finally, the French Revolution. The new identity symbol was "solidarity"—from Fourier to Durkheim. It replaced nature with moral claims. This again did not prove to be very helpful. How could one expect to control rapid social change by appeals to solidarity? Durkheim could only say that the modern division of labor would *require* a new type of solidarity, but his dissertation ends with the injunction: "En un mot, notre premier devoir actuellement est de nous faire une morale" (406). And "solidarity" has simply become a word justifying tax increases (Germany) or demonstrating public spending in rural districts (Mexico).

If morality does not do its job, we need politics as a supplement. The identity slogan of the twentieth century, always set against stratification, unfair distribution, exploitation, and suppression, became the assimilation of living conditions of the whole population by political means and in particular by democratization of the political system itself and by a politically guided economic development. This, too, did not succeed, either in Manhattan or anywhere else. It seems, therefore, that we have to prepare ourselves to live with a society that does not provide for happiness, or for solidarity, or for a desirable assimilation of living conditions. There may be occasions where people can meet and critique society, but to call this "civil society" is pure hypocrisy given the facts we have to endure.

However, we may have chosen the wrong distinction, that is, the wrong form of differentiation. The present state of world society can hardly be explained as a consequence of stratification. The dominant type of system-building within contemporary society relates to functions, not to social status, rank, and hierarchical order. The so-called "class society" was already a consequence of functional differentiation, resulting, in particular, from the differentiation of the economic and the educational subsystems of society. This completely changes the semantical possibilities of constructing the unity of the system as distinguished identity. In view of unjust stratification one could find comfort in humanistic terms, focusing on the life humans can lead in society. Under the regime of functional differentiation we can maintain similar ideas, but now it becomes a question of social policy and social work as one of the many subsystems of society (Baecker). Foucault already saw the close relation between humanistic concerns and tightly regulated and controlled policies of inclusion (Foucault; see also Bender). When the focus shifts from stratification to functional differentiation, the symbolic representation of the identity of the societal system can no longer refer to human nature.[5] In fact, the new mythology has tried to justify differentiation *as such.*[6]

Looking at attempts to define the meaning of (functional) differentiation and thereby the unity of the differentiated system we find the same trend toward increasing skepticism. The first idea was, of course, that "division of labor" would increase welfare and produce a surplus available for new investment and/or for distribution. When this idea was transferred from organizations to society and from roles to societal subsystems it became the "project modernization" after World War II. The basic idea now was coherent modernization. If only society could succeed to modernize each of its function systems—to arrange for a market economy, for democracy, for universal literacy, for free "public opinion," and for research oriented by theory and method only (and not by social convenience)—then the hidden logic of functional differentiation (or invisible hand?) would grant success, i.e., an improved society. This project (or projection) depended upon specific subdistinctions for each function system, such as market economy versus planned economy or democracy versus dictatorial regimes, but the preoccupation with these distinctions prevented the discussion of the question of why one could expect "modernized" function systems to support one another and to cooperate toward a better future. Nor did the neo-Marxist critique of modernization understand the problem, but rather turned back to a neoneohumanistic critique of class structures. But if system rationality depends upon a high degree of specialization and indifference, then how could one expect and even take for granted that "system integration" comes about? Would it not be more probable that developing systems would create difficulties, if not unsolvable problems, for each other—such as the internationalization of financial markets for any kind of socialist policy, or the welfare state for the rule of law supervised by a constitutional court (see Grimm), or microphysics (atomic energy) or biochemistry (genetic technology) for a political or legal handling of risks?

Given such problems the recent discussion replaces "integration" with "guidance" or "steering capacity" and refuses to give up the hope that such instruments would—more or less—work (see Willke). Or it simply prefers the soft language of undefined terms like "institutions," "culture," and "ethics" to maintain hope. The question, of course, is not to choose between dogmatic optimism and dogmatic pessimism. The problem, rather, is how to construct the semantical artifact, the symbolic identity of society under the given conditions we have to face at the end of this century. And if we have to do it in a "dogmatic," that is, unjustifiable way, why is the option that of optimism versus pessimism?

IV

We are now prepared to come back to our question: Why does society describe itself as postmodern?

There are several easy but superficial explanations.

(1) Intellectuals, in particular postneo-Marxist intellectuals, who have lost confidence in their own theories and want to talk about that loss, tend to generalize their fate and tend to think that everybody finds him- or herself in the same situation.

(2) Origin has been replaced with history plus reform (Conring). However, by what are we going to replace reform if we can look back at so many unsuccessful attempts (Brunsson and Olsen)? The political system seems to substitute scandals for reforms, or at least we can observe a negative correlation between less reform and more scandals.[7] In business, the incentive for innovation seems to shift from improvement in terms of accepted objectives to crisis management, to attempts to avoid the worst.[8] However, reforms relate only to organizations, whereas the discourse about postmodernity relates to much broader concerns. It may overgeneralize disappointments with organizational reforms (for instance in schools and universities), but then we need an explanation for this overgeneralization.

(3) In recent years society has produced more and more communication about its environment. This refers primarily to the ecological environment, but the bodies and minds of individuals also belong to the environment because they are not produced and reproduced by communication, but by their own biochemical, neurophysiological, or psychical operations. There are increasing doubts about whether society (that is, self-reproducing social operations) can "control" its environment changed by its own output (see Luhmann)—again, the ecological conditions and five or six billion human individuals. Demographic changes (population increase), migrations, but also the increasing tendency to immediate vio-

lence become problems of social concern. But how could we describe society when the description has to admit (or to prove the contrary) that the system cannot, in spite of tight causal couplings, adapt to its environment? Or could we handle this problem by reformulating the identity of the system?

(4) The main preoccupation of intellectuals is no longer wisdom, nor prudence, nor reason, but second-order descriptions. They describe *how* others describe what others describe. This may amount to the loose talk of French writers or to the dreadful rigor of analytic philosophers. Second-order description, however, seems to be a general characteristic of a specialization concerned with the interpretation of texts. This form of communication seems to have difficulties in producing stable "eigen-values," as Heinz von Foerster would predict *(Observing Systems)*, and "postmodernity" may offer itself as an appropriate conclusion.

Our analysis of the historical semantics of modern society adds a further and more convincing argument. Whether we have to justify differentiation or compensate for it, and whether the dominant type is stratification or functional differentiation, we find a remarkable loss of confidence in symbolic presentations of the essence, or meaning, or unity of the system. The trend begins in the seventeenth century, when the "in spite of . . ." justifications were displaced from the past to the (always uncertain) future. (The corresponding change of the meaning of "revolution" is a good indicator.) It then moves from nature (happiness) to morality (solidarity). This requires political supplements and makes visible that these did not and probably cannot succeed. And if stratification or class society is no longer the main problem, the discussion about the advantages of functional differentiation only repeats the same experience. It also moves from progressivist hopes to increasing doubts. At the end of this development we find the phantom center "civil society," that is, free associations of individuals who can talk about complaints and improvements. And this is "práxis" as self-satisfying activity, whereas poíesis or reproduction has to be done by the function systems and their organizations, which cause all the trouble.

We may call this modern or postmodern society. The question is rather whether it makes any sense to use a historical distinction to mark the problem. The distinction of before and after will not prove to be very helpful. Like the rhetorical scheme of *antiqui/moderni* in the Middle Ages and the Renaissance, it is a scheme to organize second-order descriptions (see Buck; Goesmann; and Black). It shares the weakness of all indications and distinctions discussed so far. Society can describe itself as modern or as postmodern, but if it does so, what is the information? What is the difference that makes a difference?

This question leads to some concluding remarks. In the course of our discussion we have met several distinctions such as reason and reality, or modern and postmodern, or differentiation and the unity of the differentiated system. Such distinctions allow for crossing their internal boundaries. They are "frames" for observing and describing identities. But then, we will need a theory of frames, including, as Derrida would say, a frame for the theory of frames (50). Can a distinction frame itself? But then, how to move to another distinction, how to make, in Gotthard Günther's terms, a transjunctional move? Or are we forced, by using a distinction, to forget the unity of the distinction, to leave its frame unattended?

On a very abstract level systems theory may offer a frame for a discussion of framing (see Roberts).

For systems theory the answer to this question is not difficult. (But then, why use systems theory as a frame? Only because it can give an answer to the question and apply it autologically to itself?) The frame is the self-produced and reproduced difference of the system and its environment. This production produces operational closure and thereby a form. At the inner side of the form (and only there) the system can make distinctions and thereby frame its own (but only its own!) observations. Now, the system can distinguish itself with all the consequences of a reentry of the form in the form that we have outlined at the beginning. The totality of its operations becomes unobservable, and if the system tries nevertheless to observe this totality, it becomes the victim of a totalitarian logic. The self-description of the self-intransparent system has to use the form of a paradox, a form with infinite burdens of information and it has to look for one or more distinguishable identities that "unfold" the paradox, reduce the amount of needed information, construct redundancies, and transform unconditioned into conditioned knowledge (see Krippendorff). All elaborated cognition will reduce self-created uncertainty and will only lead to contingent results. Such results may seem useful to some—and detestable to others. But this remains acceptable, because the question of the unity of the distinction always leads back to the paradox—and one can show this to others and accept it for oneself. In view of all the fine prospects offered to Doctor Johnson in Scotland, there may be only one that is really attractive—the way back to England, the way back to the origin, the way back to the paradox.

Is this, after all, a postmodern theory? Maybe, but then the adherents of postmodern conceptions will finally know what they are talking about.

Notes

1. See Habermas "Können" and *The Philosophical Discourse of Modernity*, chapter 12.

2. In systems terms, transcendental refers to consciousness and linguistic to communication, and these are but different operations to produce different types of systems.

3. See Conring. The treatise ends with proposals for improving the law (chapter 35) that could not be derived from the "origin" of the law but from the accidents of its history.

4. The doctrine of the social contract (Hobbes, Pufendorf, and others) seems to function as a substitute for the lost belief in the presence of "origins." Formally, it is a very similar construct, the irrevocable contract is the new origin of the validity of the law. But it has this position due to a legal fiction and a hidden paradox. For a contract presupposes already the binding force of the law.

5. The point needs more argument, of course. But the argument would require a previous clarification of functional systems differentiation.

6. The famous distinction of social integration and system integration came very close to this point, but it did not see the real problem and treated it as if it were a theory mistake (see Lockwood). Moreover, if there are two forms of integration, one would like to hear something about the integration of integrations, and this would require a definition of the concept of integration. If it is to include system integration it can no longer be defined as consensus.

7. The explanation may well be that the system needs a nonnatural way to create free positions.

8. Or, in Odo Marquard's terms: "Zielstreber" becomes "Defektflüchter."

Works Cited

Baecker, Dirk. "Soziale Hilfe als Funktionssystem der Gesellschaft." *Zeitschrift für Soziologie* 23 (1994): 93–110.

Bender, John. *Imagining the Penitentiary: Fiction and the Architecture of Mind in Eighteenth-Century England.* Chicago: University of Chicago Press, 1987.

Black, Robert. "Ancients and Moderns in the Renaissance: Rhetoric and History in Accolti's *Dialogue on the Preeminence of Men of His Own Time.*" *Journal of the History of Ideas* 43 (1982): 3–32.

Brunsson, Nils and Johan P. Olsen. *The Reforming Organization?* London: Routledge, 1993.

Buck, August. *Die "querelle des anciens et modernes" im italienischen Selbstverständnis der Renaissance und des Barocks.* Wiesbaden: Steiner, 1973.

Conring, Hermann. *De origine iuris germanici: Commentarius Historicus.* Helmstedt: Henning Myller, 1643.

Derrida, Jacques. *La vérité en peinture.* Paris: Flammarion, 1978.

Durkheim, Emile. *De la division du travail social.* 1893. Reprint, Paris: PUF, 1973.

Foucault, Michel. *Surveiller et punir: Naissance de la prison.* Paris: Gallimard, 1975.

Goesmann, Elisabeth. *Antiqui und Moderni im Mittelalter: Eine geschichtliche Standorthestimmung.* München: Schöningh, 1974.

Grimm, Dieter. *Die Zukunft der Vergassung.* Frankfurt: Suhrkamp, 1991.

Habermas, Jürgen. "Können komplexe Gesellschaften eine vernünftige Identität ausbilden?" In *Zwei Reden aus Anlaß der Verleihung des Hegel-Preises 1973 der Stadt Stuttgart an Jürgen Habermas am 19. Januar 1974,* edited by Jürgen Habermas and Dieter Henrich. Frankfurt: Suhrkamp, 1974.

———. *The Philosophical Discourse of Modernity: Twelve Lectures.* Translated by Frederick G. Lawrence. Cambridge: MIT Press, 1987.

Hale, Sir Matthew. *The History of the Common Law of England.* Edited by Charles M. Gray. 1713. Reprint, Chicago: University of Chicago Press, 1971.

Husserl, Edmund. "Die Krisis des europäischen Menschentums und die Philosophie." In *Die Krisis der europäischen Wissenschaften und die Transzendentale Phänomenologie.* Vol. 6 of *Husserliana.* Den Haag: Nijhoff, 1954.

Krippendorff, Klaus. "Paradox and Information." In *Progress in Communication Sciences* 5, edited by Brenda Dervin and Melvin J. Voigt. Norwood, N.J.: ABLEX, 1984.

Lockwood, David. "Social Integration and System Integration." In *Social Change: Explorations, Diagnoses, and Conjectures,* edited by George K. Zollschan and Walter Hirsch. New York: Wiley, 1976.

Löfgren, Lars. "Complexity Descriptions of Systems: A Foundational Study." *International Journal of General Systems* 3 (1977): 227–32.

Luhmann, Niklas. *Ecological Communication.* Translated by John Bednarz Jr. Chicago: University of Chicago Press, 1989.

Lyotard, Jean-François. *The Postmodern Condition: A Report on Knowledge.* Translated by Geoff Bennington and Brian Massumi. Minneapolis: University of Minnesota Press, 1984.

Marquard, Odo. "Kompensation—Überlegungen zu einer Verlaufsfigur geschichtlicher Prozeese." In *Aesthetica und Anaesthetica: Philosophische Überlegungen.* Paderborn: Schöningh, 1989.

Roberts, David. "Die Paradoxie der Form in der Literatur." In *Probleme der Form,* edited by Dirk Baecker. Frankfurt: Suhrkamp, 1993.

Ruesch, Jürgen, and Gregory Bateson. *Communication: The Social Matrix of Psychiatry.* 2d ed. New York: Norton, 1968.

Spencer Brown, George. *Laws of Form.* 2d ed. New York: Dutton, 1979.

von Foerster, Heinz. "Für Niklas Luhmann: Wie rekursiv ist Kommunikation?" *Teoria Sociologica* 1.2 (1993): 61–85.

———. *Observing Systems.* Seaside, Calif.: Intersystems, 1981.

———. "Principles of Self-Organization—In a Socio-Managerial Context." In *Self-Organization and Management of Social Systems: Insights, Promises, Doubts, and Questions,* edited by Hans Ulrich and Gilbert J. B. Probst. Berlin: Springer, 1984.

Willke, Helmut. *Ironie des Staates: Grundlinien einer Staatstheorie polyzentrischer Gesellschaft.* Frankfurt: Suhrkamp, 1992.

Winograd, Terry, and Fernando Flores. *Understanding Computers and Cognition: A New Foundation for Design.* Reading, Mass.: Addison-Wesley, 1987.

2. No Exit?
(Response to Luhmann)

Peter Uwe Hohendahl

It is not clear to me whether Niklas Luhmann's essay "Why Does Society Describe Itself as Postmodern?" has a message, yet there appears to be a thesis, which is articulated at the end. This thesis relies on the distinction between simple observation and second-order observation. Hence the answer to the paper's initial question of whether our society is a modern or a postmodern one is moved to a higher level: the distinction between a modern and a postmodern society or between modernity and post-modernity is a scheme to organize second-order descriptions. Whether we describe our present condition as modern or as postmodern depends on our viewpoint or frame of reference. It is not a matter of an objective description, which then can be called true or false.

At this point two questions have to be raised: First, why is this so? And second, if the above mentioned thesis is correct, or at least persuasive, what are its implications for the ongoing debate about postmodernity and postmodernism?

Luhmann holds that systems theory as a theory that grounds itself in difference has a distinctive advantage over conventional ways of framing the postmodernism debate. Working with the distinction between system and environment, systems theory assumes that the system arrives at a self-description in the form of a paradox that has to be "unfolded" or brought under control and made available through different perspectives. The implication is a self-created uncertainty of the system/environment distinction that one might want to call "postmodern." Hence systems theory itself is possibly the answer to the question that it means to answer: what distinguishes postmodernity from modernity? But if this is the case, the answer immediately raises another question: could systems theory give a different answer? Or, to put it differently, what happens when we leave the frame of systems theory and turn it into an object of observation and interpretation, that is, see it as a symptom of complex modern societies?

Before I turn to this question, however, I want to address the first question: why is the modernism/postmodernism distinction only a schema to organize second-order descriptions? The bulk of Luhmann's paper is actually devoted to this problem, specifically to the nature of self-referential systems as hypercomplex systems that are capable of self-observation. Luhmann offers a number of answers, most of which he finds ultimately unsatisfactory, among them the response that postmodernity is the invention of former neo-Marxist intellectuals who have become dissatisfied with their own theories or the suggestion that the discourse on postmodernity/postmodernism results from an overgeneralized critique of failed social reforms. Another suggestion is the assumption that today intellectuals are no longer concerned with wisdom or judgment, but with second-order descriptions, a situation that then could be labeled as typically postmodern. While these answers have a certain plausibility, Luhmann places the emphasis of his own reading on the broader phenomenon of *symbolic representation,* specifically the problem of the unity, respectively the meaning of the system. Differentiation makes it increasingly more difficult to talk about "Sinn."

The reasons for this difficulty are defined as historical (in contrast to a normative or transcendental explanation). They can be traced back, as Luhmann tries to show, to the seventeenth century. Of crucial importance in this context is the transition from a stratified society to a functionally differentiated society during the eighteenth century. Luhmann underscores that the problems of contemporary society are the result of a historical process in which the system reaches higher degrees of complexity and moves from a complex to a hypercomplex, i.e., self-referential system. Still, systems theory is able to describe these problems without referring to the modern/postmodern distinction; it can use a variety of frames to carry out this task. That means we have to distinguish between the description of the historical process, on the one hand, and the reading of this process under the sign of modernity and postmodernity on the other. For Luhmann, the description of the historical process is the sociologically more important task, while the problem of postmodernism is a metareflection of this task that throws an interesting light on the task itself.

I want to stay for a moment with the historical aspect, which Luhmann seems to understand as an objective, empirically provable process. Luhmann's sketch emphasizes the failures of modern society to come to grips with the results of its own development. The first failure was the idea of happiness as a response to the loss of an unquestioned stratified society, the project of the early Enlightenment. The second failure was the idea of solidarity as a response to the social problems resulting from the functional differentiation, especially during the nineteenth century (socialism). The third failure was the notion of modernization that would finally overcome the dysfunctional implications of the autonomous de- velopment of the subsystems. Consequently, Luhmann responds with con- siderable skepticism to any social and/or political project that promises to resolve the tensions and contradictions of the present system. Neither revolution nor reform are accepted as promising venues. With a certain impatience Luhmann refers to the current debate about the civil society as "pure hypocrisy" and maintains: "It seems, therefore, that we have to prepare ourselves to live with a society which does not provide for happi- ness, nor for solidarity, nor for a desirable assimilation of living con- ditions." Why is this the case? Because the systemic necessities will ulti- mately override the discussion among the individuals about possible improvements through forms of political praxis.

I want to argue that this pessimism is not a description, but rather a reading or an interpretation that depends on a specific input. In other words, Luhmann's pessimistic emphasis on the *failure* of the modern so- cietal system overgeneralizes specific features that are definitely part of the larger matrix; at the same time it underemphasizes possibilities for change through communication (public sphere). The chosen approach reinforces the dominance of the system as a complex or hypercomplex system that produces and reproduces itself (in this respect it reminds us of structuralism). It is, of course, this pessimism or skepticism that en- courages Luhmann to consider the label "postmodernism" for his find- ings. But before I unfold this aspect of the essay, I want to raise another question: is the pessimistic view a matter of a personal bias on the part of the author, or the result of the theoretical approach defined by systems theory?

It seems to me that the theoretical apparatus that Luhmann intro- duces at the beginning of the essay, i.e., a brief summary of systems theo- ry, does allow him to draw the conclusions presented at the end, but this apparatus would also allow for different emphases and possibly different conclusions.

Self-referential systems can conceive their own complexity as well as the unity of their complexity, but at the same time they are not transparent: they cannot know themselves. All they can do is use distinctions to define themselves (for instance, through the system/environment distinction). Moreover, the system has a memory function, which enables it to remember, to learn, and to correct itself through its operation. That is why the operational level is more important than the semantic level. On the operational level, hypercomplex systems define their identity, as well as modifications of their identity. It is this apparatus that allows for both the reproduction of the system and the internal criticism of its functions, for example the critique of the destructive force of an economic system that exclusively follows its own logic without concern for the environment. Thus the "failure" of the project to solve the tensions and contradictions of a functionally differentiated society (with a strong emphasis on the steering function of the economy) through "solidarity," a project commonly known as socialism, encourages either a skeptical attitude toward such a project or a motivation to rethink the problems that were left unresolved. In this process neither the resources, nor the number of possible operations are limitless, but within a self-referential system with a memory function, these resources can be strengthened and modified. "Postmodernism" as a way of organizing expectations, by the way, could be seen as a cautionary and skeptical approach to the ability of the system to address its own deficiencies (see Lyotard). The insistence on modernity as an incomplete project, on the other hand, as we find it in Habermas, could be perceived as an expression of confidence that the present problems are not unsolvable. Such an expectation should not be confused, however, with the belief that a *project* will take care of systemic problems once and for all—the mistake of all linear utopian projects without an understanding of their own implications for the structure of the system. Luhmann's personal bias comes into the foreground when he strongly expresses his preference for a skeptical attitude and his impatience with intellectuals who do not share this outlook.

I want to return now to Luhmann's definition of postmodernity and postmodernism as part of a discourse on hypercomplex systems; that is to say, I am coming back to my initial question—what are the implications of Luhmann's thesis for the ongoing postmodernism debate? Unlike the majority of critics who have embraced the term, Luhmann is not heavily invested in the concept. By distinguishing between first- and second-order descriptions, he displaces the concept to the level of the frame that enables and organizes the description. But it is understood that this particular frame can be replaced with other frames and it would take a metatheory of the frame to decide which one is more useful and fruitful for a given task. The advantage of this solution is the fact that it avoids a dogmatic commitment to one viewpoint, a position that typically results in either/or decisions (modern or postmodern).

In the beginning I suggested that systems theory itself might be involved in the quest for postmodernism. It could possibly be the answer to the question: what is postmodern? Luhmann's strategy to move the concept of postmodernism to the level of second-order observation hints at a relationship that deserves to be explored. One possible path would be the assumption that systems theory, its very structure, must be seen as a reflection of a postmodern age, in the same way that Kant and Hegel were symptomatic of the modern age. In this view, of course, systems theory becomes an object of observation and description. Systems theory's emphasis on difference and decentering can be seen as typical elements of a postmodern approach. Depending on the viewpoint, this can be regarded as a flaw or a merit. It seems, however, that the relationship of systems theory and postmodernism has to be conceptualized in a less reductive form. It would require reading systems theory as a theoretical response to the trend toward hypercomplexity, as well as part of hypercomplexity and self-referentiality. As we have seen, systems theory insists on the blind spots of the system, on the opaqueness of the system for itself, but can systems theory recognize its own blind spots or is it supposed to be the all-knowing theory machine for which opinions of people and public discussions are merely "noise" that complements its operations? To put it differently: to what extent is it possible for systems theory to view itself from the outside and reflect on the possibility of error or misconception? The claim to systematic theoretical rigor, especially the claim to a complete theoretical conception of reality, would hardly be compatible with postmodernism, which favors decentering. Hence we would acknowledge a contradiction between the structure of the theory and its historical context. But, as we have to remind ourselves, this is true only as long as we conceive of postmodernism as a semantic term for historical processes. Systems theory overcomes the before-mentioned contradiction, but also the charge of being a mere postmodern epiphenomenon by suggesting a hierarchy of framing devices that enable it to disarm the contradiction and at the same time prove its own flexibility. In making these moves, we are still within systems theory. Is there an exit? Or is every exit guarded by systems theory itself? But if this is the case, then we might be left only with the option of being inside or being without a theory at all.

3. Pre- and Post-Dialectical Materialisms: Modeling Praxis without Subjects and Objects

Marjorie Levinson

I begin with three quotations, serving as something between topic sentences and course headings.[1] One is from Marx, the other two from the more recent tradition of materialist social thought. Together, they highlight what I see as some general interests and aims governing the effort on the part of today's radical thinkers to reconceive both the practical and the categorical relations between culture and nature, the human and the nonhuman, the biological and the mechanical. These statements should also help to distinguish that project from the ecological critiques of industrial and postindustrial capitalism that develop from a conservative humanist position. I refer to writers like Jonathan Bate, who use the rhetoric of intervention to revive the primitivist, essentializing, aestheticizing, and protectionist views of nature that arose in the early nineteenth century in response to despoliations brought about by industrial capitalism, as well as to changes in consciousness promoted by that economic and social transformation.[2] Romantic period writing, many canonically definitive forms of which launch an internal critique of bourgeois competitive individualism by way of what used to be called "nature worship," has for obvious reasons become something of a resource and a touchstone for conservative ecology. It is my feeling that much of this poetry, when its reading is informed by concepts materially intimated in the technologies of the present, may release a very different picture of the human in its physical environments—or, one could say, of the physical environments that compose the human. This picture promises to be less constrained by notions of subjective priority than the models articulated by both the traditional and the revisionist readings of Romantic poetry. Rather than shore up the anthropocentric form of the subject embedded in the conservative critique of capitalism and its exploitation of natural resources, this picture could assist the general project of critique of the subject, an exercise in social transformation.

I take my first coordinate from *Dialectic of Enlightenment*. Here, Max Horkheimer and Theodor Adorno describe their task "not as the conservation of the past but as the redemption of the hopes of the past." My second heading, from Marx's *Eighteenth Brumaire*, urges the Revolution to take its poetry from the future, and not, as in 1789, from the storehouse of antiquity. Last, I repeat a parenthetical remark from T. J. Clark's essay on Clement Greenberg and the avant-garde. Questioning some classic accounts of modernism's practices of negation (specifically, its foregrounding of the physical medium in order to block bourgeois identification with and entry into the picture), Clark pauses to wonder more generally "Why, after all, *should* matter be 'resistant'? It is a modernist piety with a fairly dim ontology appended" (152–53). As usual, Clark is a model of understatement. The alleged resistance of matter could be described as more than a piety and earlier than modernism and its ontology sharpens if we conceive it in social terms, much as Clark proceeds to do in that essay. The task, as he and others conceive it, is to pin down the connections between, on the one hand, the broad range of needs entailed by particular social and economic formations, and on the other, the special experiences and ideas of the human (and of subjectivity and inwardness as its privileged forms) that meet or challenge those needs. Matter, as a trope of resistance to the human in general and to thought in particular, will turn out to have specific constitutive functions with respect to particular social formations and ideals.

The Adorno-Horkheimer distinction between conserving and redeeming the past could also be expressed as the difference between historicism and dialectics, between repetition and remembering, and between two kinds of violence: the violence of repression and the violence of reinvention. The statement also draws a line between the past and the hopes of the past, which do not lie ready to hand, empirically self-evident, *in* that past. Glossed by the caution from Marx and applied to the politics of criticism, the Frankfurt School project of redemptive historiography can be read as the confessedly impossibilist attempt to realize the hopes of the *present* rather than wait for history to redeem them. In order to do that, the critic must shape his or her practice not *to* that present but to a future that is somehow (in some coded, partial, obscure, and unselfconscious way) sealed up in contemporary material conditions. A politics or a criticism thus conceived will understand that the unreflected survey of

the present scene—the object riddled with error—cannot of itself fur-
nish a critical perspective. This is the lesson of the past.

The chief distinction made by both the Adorno-Horkheimer quotation and the slogan from Marx is between a conservative and a critical
restoration, or between what Seyla Benhabib has called a politics of ful-
fillment and a politics of transfiguration. These terms distinguish two
critical orientations that are often conflated or confused: on the one hand,
humanly liberating actions governed by ethics and agendas based on
empirical observation and designed to secure or reform existing identity
forms, and on the other, action and thought oriented toward the as yet
incompletely thinkable conditions and potentials of those given arrange-
ments and assumptions. No amount of self-inspection or sociological
analysis will yield the concepts that would organize those possibilities into
knowable forms. Since there is no breaking with the intellectual processes
of the present, what is needed is some kind of break *within* those processes, some critical opening onto their historicity.

Tim Clark's question introduces just such an opening. It raises the possibility that the seemingly axiomatic resistance of matter to mind and by
extension, nature to culture (however the content of these terms is defined), may, instead of restricting the human endowment, in fact prop it
up. As I will explain, Clark's question has implications for practices of critical knowing once nature, like the unconscious, has been subjected to a
"new and historically original penetration and colonization," such that its
"last vestiges . . . which survived on into classical capitalism are at length
eliminated" (Jameson 49). That description is Fredric Jameson's and it
helps to define what he, following Ernest Mandel, designates late or postindustrial capitalism, the situation of the Western critical practices today.

Lately, students of Romantic period writing have been looking at
styles, forms, and values long considered peripheral, epiphenomenal, or
epochally anomalous. In such literary modes as the Gothic and the sentimental, for example, readers are finding figures of subject-object, inside-
outside, self-other, intention-action, individual-group relations that do
not match up with the more familiar patterns of difference and identity,
patterns that tend to involve some sort of prolifically oppositional dynamics. What I shall call the weak forms of Romanticism (to distinguish them
from the so-called "strong" or Oedipally and dialectically organized writing of the period), when focused through the lens of the many critiques
of the subject associated with poststructuralist thinking, embody affective,
existential, economic, social, political, and even biological possibilities
toward which contemporary theory is reaching. The textualities I have

mentioned challenge the work model of activity on which the philosophies of the subject, of reflection, and of praxis are based. This model, which supports the wide range of discourses associated with the project of modernity, features the profitable transformation of nature and matter by a human (e.g., cultural, social, national) agency that is both materially empowered by this process and refined into ever increasing self-awareness and self-possession.[3] By contrast, the figures and narratives that organize a good deal of Gothic and sentimental writing do not conform to a mechanical, organic, dialectical, or deconstructive model of subject-object relations, all as it were solutions to the mind-body problem. In an essay entitled "Romantic Poetry: The State of the Art," I explore one such departure in the context of a minor Wordsworth poem, a lyrical ballad that revolves around an image and a narrative of indifference. As in the "impoverished art" of Samuel Beckett, "nothing happens, nobody comes, nobody goes" (Bersani and Dutoit). In Wordsworth's poem, however, the thematization and valorization of the negative (the mechanism of the modernist transumption of banality) does not happen either, and that is to say that the poem's rhetorical and philosophical indifference is of a different order than that which characterizes the significant insignificances of the modern. My effort in the essay is to explore the potentials of a representational practice that does not participate in economies of subjectivization and of value, economies entailed by the qualitative and philosophically founding distinction between subjects and objects.[4] In the case of the dominant varieties of Romanticism and also modernism, these economies tend to neutralize the poetry's subversive gestures.

It is my feeling that forms and effects such as these launch a second-order resistance to the dominative reason of Enlightenment, a resistance that may help us in our own struggles with various present-day forms of Enlightenment. I refer to the limits of the various critical operations, all of which feature some kind of transformative and valorizing interest in the object of study. Underlying the formal parallel is a broad, objective convergence, connecting Romanticism's fraught relation to Enlightenment, or to modernity in its first full-dress appearance, to the postmodernism that obtains in certain ways and places today, a challenge to the realizing and in effect affirmative negativity of the great modernisms, or, one could say, of the great critiques of modernity, for often these amount to the same thing. The "second-order" resistance presently discernible in certain Romantic sites today differs from the reactive, antithetical, and to that extent, formally absorbed critique embodied in Romantic nature worship, the program that Bate and other conservative critics hope to revive. In other words, both movements, the Romantic and the postmodern, can be seen as connected-in-difference to the modernity they negate. At the same time, or by another reading, both intimate a critique that is *not* so embedded. Both, at certain moments or under certain conditions, break free not just of analytic and skeptical reason, but also of the more fundamental subject-object problematic and its dialectical overcomings, symbolic fusions, and dialogic reciprocities. Both manage to refigure mind-matter, self-other, human-natural ratios by way of, in the case of Romanticism, pre- or non-Cartesian paradigms, and in postmodern critique, by embracing the category transgressions entailed by the new sciences and technologies. Both contain practices of difference or apartness (as opposed to negation, opposition, intervention) that avoid bringing forth through their confrontational coherence another and yet more total humanism.[5]

The postmodernist exercise in transfigurative thinking follows from widely experienced difficulties in continuing to perform a subjectivity that is externally bounded by hierarchies of identity-difference ratios, and that is internally structured, stratified, and driven by conflict arising from contradictions between purpose and instinct and desire and need, both of those duos reflecting the master binary of self and other. The reasons for this felt deconstruction of the classical as well as dialectical categories of difference and identity are too many and too various to recount here. Let me cite one commonly explored phenomenon—the absorption of the political and the ideological by the economic, or the chiasmic relationship between what were once conceived as distinct domains related to each other by a linear, mechanical, or reflective causality—to signal the size and complexity of the changes involved.

The change that interests me both for its own sake and for its capacity to mobilize otherwise inert strains in Romantic period writing is the changed function of nature in the present. By function, I mean action potentials, drawing on Vladimir Propp's early structuralist distinction between function and content in narrative economies. The assumed structural, material, and even ontological otherness of nature was the enabling condition for that model of the self and of the human conventionally traced to the Enlightenment and its philosophic ancestors (Bacon, Locke, and Newton). It can be seen, however, that nature's resistance to the human took a human form. In its way of asserting its otherness, it respected modes of action, opposition, and self-definition associated with the human community. Something basic in that picture and experience of the self and of humanness must change once nature begins to demonstrate a distinctive kind of agency, one that formally departs from the modalities of impulse, action, and effectivity associated with the human or cultural context. I refer to action forms that do not prolifically oppose the human in the ways familiar to us from the Hegelian, Romantic, Marxist, and also Freudian accounts. Nature is no longer that substantial resistance invoked by Kant in his preface to the *Critique of Pure Reason,* lacking which the dove of thought could not take flight. A growing number of biological and physical processes (such as weather anomalies, new diseases and epidemiological behaviors, genetic mutations) reveal a randomness (and often, an imponderable mixture of randomness and determinisms) that the available constructs of entity and environment, chance and necessity, organ and system, and even time and space cannot conceptually seize, much less control. Chaos and complexity theories represent one kind of effort to frame these nonrational, nonlinear, and irreversible patterns of change.[6] The keen and surprisingly widespread interest in these theories shown by humanists of many stripes suggests some general dissatisfaction with mainstream explanatory paradigms, and also, related to this, a general sense that the natural world has changed both in a substantive way and relationally to the human and social worlds over the last decade.[7] These natural actions do not add up to an equal and opposite subject-form, the sort of monolithic and either deified or demonized otherness that once (in the age of belief that, according to Horkheimer and Adorno, was already the beginning of Enlightenment) called forth the mythic identities of gods and heroes. Rather, these freakish natural behaviors suggest a *mutation* of agency and this puts our own agency as well as our concepts of it at risk. The boundaries between the human and the natural, the biological and the physical, the organism and the machine, the mind and the body, are now, at strategic points, breached. A degree of self-deconstruction, betraying the interdependence or imbrication of the received categories, seems to have occurred at the level of technology and scientific practice.

Finally, nature seems for the first time ever a finite domain that we are well on our way to exhausting. Like the other changes I have mentioned, this one challenges classical models of the human in a deep and qualitative way. Lacking an irreducible and as it were, self-perpetuating otherness in nature, structurally guaranteeing the ongoing recognition of the human, our transformative encounters with the physical environment cannot do the subject-making work they once did. They cannot yield the same dividends.

My response to this situation is nearly the opposite of those who advocate revival of the nature worship that marked the earliest responses to industrial capitalism. Rather than seek to conserve or restore the past, I would like to imagine what a "redemption of the hopes of the past" might look like. The plan is to return to Romanticism through the western gate, through "the poetry of the future," or through a postmodern figure of nature and thus culture, of the other and thus the self, that looks nothing like Romanticism's high arguments, but very much like some of its more retiring representational effects, or what I termed above the weak forms of Romantic period writing.

Even as I affirm this interest and urge others to pursue it, let me qualify it by stating the obvious: namely, that one would be mad not to be terrified by the changes in the structure and behaviors of the environment. In a more personal vein, I must also confess that I for one very much miss the assured interiority I remember from my own past and from the pasts available to me through many works of literature. But I also believe in the historicity of the choices available to us and in the dependence of critique on the real conditions of physical and social life and their modes of reproduction. Romanticism's discourse of nature had a critical, a utopian, and a transfigurative value in its own day, but it will not work the same magic two hundred years later and within a cultural formation that is not dominated by the commodity form, not sustained by colonial expansion, not defined by the reorganization of agrarian labor into the patterns of industrial manufacture, and not faced with a nature that patiently abides our actions and gives predictable returns on our investments.

For me, then, it is not a question of *deciding* to conduct a presentist reading of the past, nor is presentism a matter of relevant topics and remedial values. It is a question, rather, of the forms that define knowledge, objects, and experience in the present. One returns to Romanticism—or one undertakes to redeem the hopes of the past—because the pre-Enlightenment imagination students of the period are now finding in that body of writing gives a concretely sensuous and in some ways more advanced form to the post-Enlightenment stirrings and strivings that characterize the present scene.[8]

The two discourses that have structured my efforts to articulate Romanticism's postanthropological, postdialectical perspectives are Spinoza and some work in the field of theoretical biology. In the context of Romanticism's philosophical critique of Enlightenment, Spinoza provided a theory of knowledge not implicated in the Cartesian relations that defined the age's normal science, nor was this theory consistent with the

Kantian and Hegelian structures (troping particular economic and social forms) that governed the age's approved oppositional modes, its licensed subversions. I invoke Spinoza to signify a mode of representation not based on rupture, scission, or negation, and not subject to reappropriation. Not, that is, organized along the lines of material production and reproduction in their classic agrarian, industrial, and sexual (patriarchal) forms.

The unique place Spinoza holds in the history of philosophy can be traced to his postulate of a reality that is one substance, given in, or as the infinite attributes of, mind and matter, thought and extension. This assault on Cartesian dualism breaks proleptically with all the familiar idealist and materialist philosophies with their "for itself" of thought, their "in itself" of sheerly existent material reality, and the difference-in-identity of their reflective or dialectical synthesis. Spinoza builds a universe that is nothing but the thought that is god or nature ("*Deus sive natura*"). His account acknowledges the reality and the force of material and historical conditions along with the whole realm of imagination or ideology, or empirical self-evidence. At the same time, he advances the claims of a critical reason that calls such knowledge into question. The authority of this reason does not, however, derive from either external or internal referents, as in, respectively, correspondence and constructivist epistemologies. Its truth is strictly a function of the coherence, complexity, and combinatory power of its articulation and its products. In brief, what Spinoza offers is a nondualistic but nonreductive materialism, very different from the Kantian analytic, which negotiates the subject-object split by positing the subjective constitution of experience and the objective regulation of the subject via the categories and the transcendental time-space intuitions. Spinoza's double-aspect monism also differs from the Hegelian *aufhebung* which, like Kant's analytic, draws the objective term into the dialectic by rewriting it as a displaced, disguised, or undeveloped form of subjectivity (or Spirit). For Spinoza, nature is not the delimiting and thus instrumental negation of the human, nor is its otherness a mere ideological illusion, masking the contentious realities of social practice. (That is a view that effectively dissolves difference into an artifact of human activity, where the human amounts to an essentialist postulate of self-realizing activity.) Spinoza's argument for the immanence of knowledge to its object—his proposition that the mind is nothing but the idea of the body, itself a modification of that larger body that is nature as a whole—stands behind Althusser's argument for scientific knowledge of a system as one possible product of and element in that system (and thus, as both truth and illusion at the same time, or as Spinoza might say, under different aspects and at different levels).

Interestingly, Gilles Deleuze claims Spinoza as the ancestor of his postrationalist (that is, nonlogocentric) theory of excess, affects, and speed (as opposed to containment, thought, and structure) as defining formal properties. As read by Deleuze, Spinoza provides a grammar for articulating a subjectless thinking and a theory of affects that displaces traditional ethics. Affect is traced to the combinatory energy within and between individual entities, and "goodness," like "joy," is defined as "a matter of dynamism, power, and the composition of powers" (*Spinoza* 23).

Spinoza is also named by the deep ecology movement as its philosophical source. This movement takes issue with the anthropocentrism of the various protectionist or "stewardship" approaches to the environment. For deep ecology, protection is nothing more than long-range and displaced production. Human stewardship objectifies nature in the sense of converting it conceptually if not literally into "resources" that are valorized by reference to long-term availability for development of a material or spiritual kind. Deep ecology considers these two modes of development equally exploitative—their imagination of the human, moreover, no less restrictive than their concept of nature. Deep ecology is curious to imagine entities, or what Gregory Bateson calls "units of purpose," in terms of complexity, coherence, aggregative capacity, and energy, rather than structures, boundaries, linear causalities, and intentions.

This is the point of intersection with some studies in theoretical biology. In the work of Francisco Varela and Humberto Maturana, Richard Dawkins, Rupert Sheldrake, and from a different angle, Donna Haraway, one finds a rejection of the mind-matter, culture-nature ratios developed by the critical philosophies, reflection theories, and philosophies of praxis, all of which define nature, however socially produced, as the bounding outline to the human.

Richard Dawkins, for example, redefines the meaningful biological entity as the DNA material, shifting the emphasis from the individual organism and the cell. He reconceives natural selection at this level and by reference to what he calls "the selfish gene," whose only drive is to launch itself into the next generation through any body that will get it there. In effect, Dawkins takes Darwin's critique of fixed and ahistorical species to the next, or rather more basic, level. Pursuing the implications of this shift, Dawkins considers the external products or behavioral effects of the discrete and bounded organism as, in a strict sense, its phenotypic expression. Many of these so-called "animal artifacts" (e.g., termite mounds, beehives) are collectively produced. Often, the producing community is one whose genetic material or that part of it relevant to the artifact is distributed among many discrete bodies in a fashion no different from the distribution of genetic material among organs and systems within individual bodies. In tandem, then, with his challenge to traditional notions of phenotypic integrity, Dawkins draws on the parallel cited above to extend his description of the *genotype* beyond the classically defined individual organism. Further, through an enlarged but conceptually conservative description of the parasite-host relationship, Dawkins challenges the notion of genetic purity within the individual body and its organs and cells. Finally, he disputes the accepted distinction between growth and reproduction, calling the question on individual entities in a diachronic way, or with respect to discrete generations.

Maturana and Varela challenge the traditional distinction between context and organism by defining cognition not as "a grasping of an external reality, but as the specification of one."[9] In a way that seems contrary to Dawkins's extension of the geno- and phenotype into the environment but that results in a comparable dismantling of the received binaries, these authors conceive discrete systems within and between organisms as both autonomous and recursive, triggered by the environment to release internally determined activity. The autonomy of the system, thus, is that of a *composite* unity. It is realized through neighborhood relations, one component of which is the system's behavioral or performative specifications of that unity. This is the meaning of "recursive" as used by these authors. The question they put is not how does the organism obtain information about its environment, but how does it happen that the organism has the structure that permits it to specify (i.e., operate effectively in) the medium in which it exists. This amounts to a question of representation as survival, or "autopoiesis," a word they coin. The organization of an autopoietic system is nothing but its domain of interactions and this domain can survive the change of every single property and element in the system. The ghost in the rationalist and materialist machine is shown to be nothing but the survival of a form of organization in the real world as perceived by another system in that world, a system perforce specified by that organization and thus included in it. This is a radically simple and, for most of us, counterintuitive way to conceive the identity of living systems, of the relations within and between them, and between system and environment.

Rupert Sheldrake proposes the concept of morphic fields organized by morphic resonance as a way of overcoming another kind of dualism inaugurated by modern science, namely, its recognition of evolution based on genetic memory for the biological kingdom, but not for the physical universe. Sheldrake tries at once to liberate the concept of formative memory from the spatially bounded organism (positing something very like action at a distance and across temporal divides) and to propose for the physical and, by convention, inanimate universe "habits" or predispositions to behavior based on past behavior patterns, rather than on eternal and immutable laws. Morphic fields, like the known fields of physics, are nonmaterial regions of influence extending in space and continuing in time. They are localized in and around the systems they organize. According to Sheldrake, they are the reason why we can speak rigorously of an evolutionary and historical universe governed by probabilities influenced by past events, all of which develop within nature and history. The structure of the fields within which organisms and the physical world develop depends on what has happened before. They represent a kind of pooled or collective memory of the species, based on the mechanism of morphic resonance, itself based on similarity: "The more similar an organism to previous organisms, the greater their influence on it by morphic resonance" (Sheldrake 108). Unlike the familiar kinds of field-influence, however, there is no actual transfer of energy. Sheldrake describes morphogenetic fields as probability structures in which the influence of the most common past types combines to increase the probability that such types will occur again.

The anti-instrumentalist and literally poststructuralist models of nature advanced by these writers draw on the observation common to a range of sciences that the boundaries between the human and the natural, the animate and inanimate, are weakening, thereby undoing the defining closures of those binary terms. Bill McKibben, an environmental journalist, calls this state of affairs the *second* end to nature. According to McKibben, nature came to its first end as far back as the 1930s, with the ecological disaster of a pollution reaching right into the basic physical conditions of human life. The second apocalypse comes into being with the profound ontological changes suggested and in some cases already realized by the new reproductive technologies, such that the very laws whereby the biological and physical worlds perpetuate themselves can be altered by genetic engineering. It is not just that we know how to make new things and new classes of things. Rather, we have developed modes of production capable of dissolving the classical groundplot of making and of self-making, of objects and of subjects, other and self, matter and mind, nature and culture.

This development, its potential for abuse terrible beyond all imagining, is a material fact of our moment and it is the most epochally specific fact of our times. For that reason alone, the descriptive, projective, and theory accounts that try to factor it in (the work of Donna Haraway is exemplary of this movement) are an improvement on the wishful thinking built into many of our forms of intellectual production, especially those cultural critiques that put the question of otherness and appropriation. If there is any descriptive value left in Marxism's structural analysis, it may be here that it is seen: namely, in the emergence of a mode of production that incorporates and surpasses biological reproduction. To be sure, this infrastructural element coexists with historically residual but still robust modes of production (for example, industrial, monopoly, and semiotic or simulacral capitalism). What we are perhaps witnessing is something comparable to the untidy but in retrospect sharply revolutionary emergence of the commodity form, that philosopher's stone that turns labor and use-value, human histories of making and doing, into petrified things that paradoxically immortalize the living value that they ceaselessly consume. In the strange world of deep environmental pollution and genetic engineering, we have made technology and its byproducts immanent in the natural world in a most literal way. In effect, we have undermined the very concept of raw material, not just by reference to histories of the social production of nature, but by altering the structure of biological forms and processes. Could this set the material conditions for the collapse or surpassing of the subject-object problematic, an end that is also in some way a return to a pre-Enlightenment episteme—cyborgs converging on gargoyles?

That is a coupling that sends us back, via Fredric Jameson, to Marx's injunction against the taking of moral positions, and which underlines yet again the difference between a conservative and a redemptive use of nature and the past. Jameson urges materialist critique to do the impossible: to think the cultural evolution of capitalism dialectically, as catastrophe and progress, baleful and liberating, all together (47). In the context under discussion here, the task would involve searching out ways to reinvent value, intention, production, and even survival in the absence of that relational identity that the human had enjoyed through its assured engagements with a nature that symbolized *"das ganz Andere,"* the entirely Other. Paradoxically, the deepest assault on the human may be the experience of its unboundedness and unstoppability. Quantity changes quality; in a world where the human is everywhere, how can reappropriation, the action classically constitutive of humanness and its effects, proceed? From what site would it proceed and what body, what boundaries, would this process enlarge? The second end to nature is also the second end to man, one that makes the Foucauldian farewell look like another myth of Enlightenment.

Notes

1. This essay was written for a public debate with Jonathan Bate, author of *Romantic Ecology: Wordsworth and the Environmental Tradition.* I have revised the essay to liberate it from that polemical context, but the broad strokes and the telegraphic style of the essay remain unchanged.

2. Bate wants to recover Romanticism's antithetical critique of Enlightenment, its argument for an Idea of nature that opposes the utilitarian, commercial, and progressivist values and tendencies of the age, or what we might call its economic and cultural dominants, both of them organized around the commodity form. This argument has stood demystified, or at the least heavily qualified, ever since Geoffrey Hartman exposed the Hegelian but nontriumphalist groundplot of Wordsworth's poetry, its negative dialectics. Bate justifies his attempt to revive what is in essence M. H. Abrams's natural-supernaturalist argument in several, deeply questionable ways that are not worth disputing here. The strongest and most sustained defense Bate offers centers on the present environmental crisis. That phrase describes a situation in which the most basic resources and conditions of human life on this planet (or more modestly, of the social organization of human life that has become normative over the past sixty years) will probably be exhausted or irreversibly contaminated in the lifetime of persons now alive. In other words, the conditions are set for a practical transcendence of regional, classist, national, and even economic self-interest. Clearly, the argument runs, this is the moment to usher back in the most universalizing claims of Romantic nature worship, with its advocacy of a reverential stewardship of the environment as the distinctively, essentially, and ennoblingly human attitude.

For Bate, today's world badly needs the deep-structure limits within which Romanticism's acts of mind took place. Through the revisionist readings of the past decade and the skeptical or deconstructive work that preceded it, we have learned how those consecrating acts, despite their humanizing intentions and effects, ultimately reinforced the ontological difference between nature and mind in order to confirm the latter—and a distinctively productivist form of the latter—in its scope and priority. This is exactly the effect Bate hopes to recover.

For me, the category slippage that seems so widespread and definitive an experience of life in our times has no choice but to move forward into a yet more dangerously blended and labile future. My interest in Romanticism is the opposite of Bate's.

3. The processes of value production as articulated through Hegelian, Marxist, and also Freudian theory may be read as confirming a specifically gendered construct of sexual reproduction. I refer to the way in which the dynamics of self-enriching alienation recapitulate an allegory of insemination. A substance that is figured as essential and definitive because the generative human element is alienated from its source, incarnated through its mixture with an ontologically "other" substance (that is to say, the ahistorical, as it were, given material body of the woman), and reappropriated in its developed, valorized state by the original male agent, with the twofold effect of enlarging and enriching the male body and humanizing, in the sense of conferring a more realized form upon, the female. In light of this homology, the biological creativity of woman—as close to a universal stereotype as one gets—may be read as an ideologically pressured denial of the primary genetic productivity assigned to men.

4. One postmodern figure for a nondialectical model of difference and identity is Deleuze's "fold." Instead of a subject and an object, an inside and an outside when these are conceived as structurally distinct and (however infinitesimally) separated domains, the fold allows us to think differentiation, orientation, position, and therefore identity in terms of topological variation: not objects and events, but ceaseless self-relation. "The outside is not a fixed limit but a moving matter animated by peristaltic movements, folds and foldings that together make up an inside: they are not something other than the outside, but precisely the inside of the outside." Invagination, chiasmus, and the more traditional Möbius strip are metaphors that belong under this concept (*Foucault* 96, 97).

5. There are overlaps here with strains of feminist work in Romantic studies that focus on writing that does not share the transformational and valorizing ambitions of the canonical verse. One thinks, too, of Paul Hamilton's current study of literalism in Romantic writing, a zero-degree discourse that is documentary without being mimetic, exemplifying a performative poetics of which he finds instances in Dorothy Wordsworth's journals. Or there is Alan Liu's work on detail that does not accumulate into picture or design but that remains extravagant, excessive, ornamental, redundant, erotic, nonrelational.

Another related project is the effort in postcolonial studies to articulate cultural otherness without in the same stroke assimilating it by orientalizing it, the end result being the subordination of the other to the privileged identity-term of the system in question. One example of this would be Ashis Nandy's construction of a critical traditionalism drawing on premodern cultural practices as well as a theorized resistance politics: as it were, pairing Gandhi and Gramsci. See Nandy's *The Intimate Enemy*.

6. See the recent discussions of complexity theory by Lewin, Waldrop, and Prigogine and Stengers.

7. In addition, many of the new imaging technologies (MRI for example) are prompting a reappraisal of anatomical structures and structuration processes, such that topology, rather than surface-depth, exterior-interior relations, provides the cognitive schema.

8. See, for example, Kroker, Sloterdijk, Taussig, and Mann.

9. Maturana and Varela, xv. See also Wolfe for a discussion of the bearing of autopoiesis on Niklas Luhmann's systems theory. Also, for a generous sampling of new mind-body paradigm exploration, see Crary and Kwinter.

Works Cited

Bate, Jonathan. *Romantic Ecology: Wordsworth and the Environmental Tradition.* London: Routledge, 1991.

Bateson, Gregory. *Steps to an Ecology of Mind.* New York: Ballantine, 1972.

Benhabib, Seyla. *Critique, Norm, and Utopia: A Study of the Foundations of Critical Theory.* New York: Columbia University Press, 1986.

Bersani, Leo, and Ulysse Dutoit. *Arts of Impoverishment: Beckett, Rothko, Resnais.* Cambridge: Harvard University Press, 1993.

Clark, T. J. "Clement Greenberg's Theory of Art." *Critical Inquiry* 9, no. 1 (fall 1982): 139–56.

Crary, Jonathan, and Sanford Kwinter, eds. *Incorporations.* New York: Zone, 1992.

Dawkins, Richard. *The Extended Phenotype.* Oxford: Oxford University Press, 1982.

——. *The Selfish Gene.* Oxford: Oxford University Press, 1976.

Deleuze, Gilles. *Foucault.* Translated by Seán Hand. Minneapolis: University of Minnesota Press, 1988.

——. *Spinoza: Practical Philosophy.* Translated by Robert Hurley. San Francisco: City Lights, 1988.

Haraway, Donna J. *Simians, Cyborgs, and Women: The Reinvention of Nature.* New York: Routledge, 1991.

Horkheimer, Max, and Theodor Adorno. *Dialectic of Enlightenment.* Translated by John Cumming. New York: Continuum, 1986.

Jameson, Fredric. *Postmodernism; or, The Cultural Logic of Late Capitalism.* Durham, N.C.: Duke University Press, 1991.

Kroker, Arthur. *The Possessed Individual: Technology and the French Postmodern.* New York: St. Martin's, 1992.

Levinson, Marjorie. "Romantic Poetry: The State of the Art." *MLQ* 54, no. 2 (June 1993): 183–214.

Lewin, Roger. *Complexity: Life at the Edge of Chaos.* New York: Macmillan, 1992.

Mann, Paul. *The Theory-Death of the Avant-Garde.* Bloomington: Indiana University Press, 1991.

Marx, Karl. *The Eighteenth Brumaire of Louis Bonaparte.* New York: International, 1963.

Maturana, Humberto, and Francisco Varela. *Autopoiesis and Cognition: The Realization of the Living.* Dordrecht and Boston: D. Reidel, 1980.

McKibben, Bill. *The End of Nature.* New York: Doubleday, 1989.

Nandy, Ashis. *The Intimate Enemy: Loss and Recovery of Self under Colonialism.* Delhi and Oxford: Oxford University Press, 1983.

Prigogine, Ilya, and Isabelle Stengers. *Order Out of Chaos: Man's New Dialogue with Nature.* New York: Bantam, 1984.

Sheldrake, Rupert. *The Presence of the Past: Morphic Resonance and the Habits of Nature.* New York: Random House, 1988.

Sloterdijk, Peter. *Critique of Cynical Reason.* Translated by Michael Eldred. Minneapolis: University of Minnesota Press, 1987.

Taussig, Michael. *The Nervous System.* New York: Routledge, 1992.

Waldrop, M. Mitchell. *Complexity: The Emerging Science at the Edge of Order and Chaos.* New York: Simon and Schuster, 1992.

Wolfe, Cary. "Making Contingency Safe for Liberalism: The Pragmatics of Epistemology in Rorty and Luhmann." *New German Critique* 61 (winter 1994): 101–27.

4. Immanent Systems, Transcendental Temptations, and the Limits of Ethics

William Rasch

Perhaps modernity—as the name given to an obsessive process of self-description—should describe itself yet again as a force field of competing anxieties. We have become distinctly suspicious of transcendental attempts to construct inviolate and panoramic levels of vision labeled God, Reason, or Truth. Yet, because of political or moral commitments, we are equally disinclined to relinquish "critical" perspectives from which we presume not only to see the world as it is, but also to utter judgments about its inadequacies. From their mid-century vantage point, Horkheimer and Adorno found themselves uneasily negotiating this terrain. According to them, the "pure immanence of positivism," which they described as the "ultimate product" of Enlightenment, was driven by a fear of the outside. "Nothing at all may remain outside, because the mere idea of outsideness is the very source of fear" (16). In the years immediately surrounding the 1969 republication of *Dialectic of Enlightenment*, critical theory waged war against positivism in the name of an outside that was seen as the (at least utopic) other of the all-pervasive administered society, an other that may inhere in the cracks and fissures of immanence, but an other that nevertheless remains outside of the Same. The fear that motivated Adorno and his compatriots was not the fear of the outside they had attributed to positivism, but rather a fear of the *loss* of the outside, a fear that lingers in much of what calls itself postmodern. Yet, Enlightenment, once loosed on the world, can apparently never be denied, for in the decades that followed, a new "ultimate product" has appeared on the scene—a revised and revitalized systems theory in the life and social sciences, whose immanence and whose evacuation of the outside promises to be even more radical and complete.

According to systems theory sociologist Niklas Luhmann, the traditional question of access to or knowledge of the outside has forced itself upon us again in the twentieth century, paradoxically, as the result of empirical research in the various sciences. As examples he cites the inevitable self-referential aspects of quantum physics ("the best-known example"), linguistics ("the fact that research into language has to make use of language"), the sociology of knowledge ("which had demonstrated at least the influence of social factors on all knowledge," including, of necessity, "this statement itself"), and, perhaps most significantly, cognitive science:

> Brain research has shown that the brain is not able to maintain any contact with the outer world on the level of its own operations, but—from the perspective of information—operates closed in upon itself. This is obviously also true for the brains of those engaged in brain research. ("Cognitive Program" 64)

It would seem, then, that what the disciplines claim to have discovered about the world has made it exceedingly difficult—indeed, has made it impossible—for them to say that they can in fact discover anything about the world. Luhmann concludes, therefore, that

> knowing is only a self-referential process. Knowledge can only know itself, although it can—as if out of the corner of its eye—determine that this is only possible if there is more than only cognition. Cognition deals with an external world that remains unknown and has to, as a result, come to see that it cannot see what it cannot see. ("Cognitive Program" 65)

That Luhmann emphasizes "brain research" is no accident. He refers here to the work of the biologist Humberto Maturana, who developed the notion of autopoiesis—the self-reproduction of a system's network of elements from that very same network of elements—to describe the essential feature of living systems. According to Maturana's own accounts, the necessity of defining living systems as operationally closed arose from efforts to describe the activities of the nervous system in light of empirical experiments dealing with visual perception, especially the perception of color.[1] The experimental and experiential evidence of frogs, pigeons, and humans led Maturana (first alone, then later with his colleague Francisco Varela) to conclude that the nervous system as such cannot distinguish between illusion, hallucination, and perception. Such a distinction can only be made retrospectively, "through the use of a different experience as a meta-experiential authoritative criterion of distinction"

(Maturana, "Biological Foundations," 55). In other words, one can only affirm an optical illusion by reference to some other standard (touch, say), which is then constructed as authoritative. Therefore, bucking the orthodoxies of the 1950s and 1960s, which viewed perception in terms of representations of the outside world or as informational "inputs" into a system open to its environment, Maturana defined the nervous system as operationally closed, autonomous, and self-referential:

> [A]ll that is accessible to the nervous system at any point are states of rela-
> tive activity holding between nerve cells, and all that to which any given
> state of relative activity can give rise are further states of relative activity in
> other nerve cells by *forming* those states of relative activity to which they
> respond. (*Autopoiesis* 22)

Thus, "[t]he relations with which the nervous system interacts are relations given by the physical interactions of the organism," and what the nervous system can be said to "represent" are "the relations given at the sensory surfaces by the interaction of the organism, and not an independent medium, least of all a description of an environment" (*Autopoiesis* 23).

As a result, all "communication" between system and environment is blocked. On the one hand, the environment can have no direct causal relationship with a system. All changes in a system are internally determined; the environment merely serves as a "triggering" device, a "perturbation" that is the catalyst for internal activity, but not the determining factor of how that activity takes shape. In like manner, a living system has no access to its environment. What it presents to itself as the outside world are representations of its own internal states. Though Maturana's model of living systems shares much with early cybernetics and systems theory (especially the notion of circular or recursive organization), it differs radically in this claim to operational closure. There simply are no informational exchanges, no informational input-output relations between autopoietic living systems and their environments.

Luhmann has appropriated and generalized Maturana's concepts of autopoiesis and operational closure in his effort to formulate a general theory of modern society as the functional differentiation of autonomous social systems.[2] There is, according to Luhmann, no *causal* relationship between environment and autopoietic social system, just as there is none between environment and living system. Social systems receive no informational inputs, no directives, no instructions, and no programs from their environments. They can be "perturbed," they can react to these "perturbations," but these "perturbations" do *not* enter the system as "units of information" that can dictate the way a system organizes its own reactions. Therefore, systems have no direct access to their environments, cannot "refer" to their environments, and can make no representation of that which is external to them. The problem systems are faced with, then, is not one of adaptation and adequacy, but rather of how the tautology of self-reference can be interrupted and unfolded in a productive manner. They are faced with the interesting and circular problem of generating "meaningful" external references where none exist.

Luhmann considers this "loss" of reference, or "loss" of the outside, to be a defining feature of the modernity we find ourselves in, and as such it makes no sense to condemn it. The task of social theory, he maintains, is not to wish for an alternate universe, but to account for the social aspects of the one we inhabit. In Adorno's nightmare vision, on the other hand, this seemingly complacent and aggressively *descriptive* articulation of modernity as the proliferation of operationally closed and functionally differentiated social systems can only be seen as the crushing victory of administered society from which there is no escape, not even an aesthetically pleasing, utopic peephole peeking out from the cell walls. Of course, framed in this way, the issue moves beyond considerations of epistemology. When access to the outside is "lost," it is generally mourned, and mourning attempts to invest this lost outside, this all but present absence, with a moral force that wants to make us "feel" environmental perturbations in the same way we once "heard" the voice of God, the traditional source of moral and political authority. If, however, moral codes (commandments), holy scripture, papal and royal edicts, and the voice of the prophets and visionaries no longer deliver direct evidence of the transcendent realm, but rather become historicized and seen as socially constructed artifacts, the task of reclaiming authority must be negotiated within the domain of an immanence that has been loosed from its transcendent anchorage. In what follows, the picture of a world that has no access to its outside is provided by Wittgenstein. In Drucilla Cornell's Derridean articulation of a possible quasi transcendence, we have the attempt to reclaim a viable outside through memory and mourning. Finally, in Luhmann we have a vision of an immanent world asserting its own authority by way of self-description.

I

"Ethics is transcendental." This, at any rate, according to Wittgenstein (*Tractatus* 6.421). His is the most straightforward articulation of the absolute inarticulateness of ethics. In the *Tractatus,* the world is a closed system. It is as it is. Since "[a]ll propositions are of equal value" (6.4), the world, in itself, neither has sense nor value. Put another way, since everything in the world is contingent, nothing in the world can express lack of contingency. Whatever guarantees the noncontingency of the world, as opposed to the contingency of the "facts" within the world, must lie outside the world, or else it too would be contingent and incapable of guaranteeing noncontingency. Absolute value is absolutely different and distant from the world. That which can be articulated in the world can make no sense of that which eludes the world, but that which eludes the world makes for the possibility of sense. There can be no communication between the mundane system of sense-making and its extra-mundane, "senseless" environment. For these reasons, then, there can be no ethical propositions. Ethics cannot be articulated, cannot deal with the world, and cannot leave describable evidence of itself in the world; it serves as the unspeakable limit or condition of the world. "Ethics is transcendental."

In "A Lecture on Ethics," Wittgenstein expresses the relationship of contingency to determinateness in terms of relative versus absolute value. A judgment of relative value is not really a judgment of value at all, but a mere statement of fact (i.e., a "good runner" is simply a person who "runs a certain number of miles in a certain number of minutes"), and "no statement of fact can ever be, or imply, a judgment of absolute value" ("Ethics" 5–6). Wittgenstein explains: Suppose it were possible to include a description of the entire world—"all the movements of all the bodies in the world dead or alive" and "all the states of mind of all human beings that ever lived"—in a "world-book." Such a book might contain all the facts of the world, but it would contain no ethical propositions. Remember the *Tractatus*: "All propositions are of equal value." That means that even the description of a "murder with all its details physical and psychological" would be on "exactly the same level as any other event, for instance the falling of a stone" ("Ethics" 6). The world is as it is means that the world is this way, not that way. It also means that at any given instance, the world *could* be *that* way, and not this way. Within the world, within the book that is the description of the world, preference for these statements of facts over those statements of facts can only be uttered by statements that themselves are chosen from a set of equal possibilities. There is no absolute preference, no necessity for choosing this set over that, and since necessity is the mother of ethics, there is no ethics uttered in the world-book.

Could there, however, be a book on ethics separate from and other than the hypothetical world-book? Wittgenstein answers: "I can only describe my feeling by the metaphor, that, if a man could write a book on Ethics which really was a book on Ethics, this book would, with an explosion, destroy all the other books in the world" ("Ethics" 7). A book on ethics could not just be one of many books in the world that could have been written otherwise. It could not sit on a shelf of books that, in fact, *have* been written otherwise. Neither could such a book sit on a shelf by itself as something unique, as a one-of-a-kind event. It would have to sit outside of the world of books that sit on shelves, outside of the world represented by the representations found on the pages of the books in the world, all of which could have been written otherwise. Therefore, a book on ethics would have to obliterate the books of the contingent world, including the book in which a book on ethics could be described. It would have to exist outside of the world that desired a book on ethics for it to be a book on ethics and not a book that desired a book on ethics. A book on ethics would have to be a book outside of the world of language, since there is no language in which a book that could not be written otherwise could be written.

But of course no *book,* in any *meaningful* sense of the word, could be written outside of the world of language. The paradoxical twistings and turnings of such statements are not meant to posit propositional truths, but rather to reveal the basic experience one has when confronted with the impossible task of ethics. At best, one can attempt to describe this experience. "I believe the best way of describing it is to say that when I have it *I wonder at the existence of the world*" ("Ethics" 8). I do not wonder that the world is as it is, for that would ensnare me in the web of contingency. Rather, I ask the quintessential metaphysical question: Why is there something rather than nothing? I could also be "tempted to say that the right expression in language for the miracle of the existence of the world, though it is not any proposition *in* language, is the existence of language itself." I have thereby shifted the "expression of the miraculous from an expression *by means of* language to the expression *by the* existence of language" ("Ethics" 11). But of course, in so expressing it, I have recaptured the miraculous in language, and in language, the miraculous ceases to be miraculous. The entire project is fraught with paradox. If ethics can only exist in the transcendental realm of necessity, then ethics can never be glimpsed from within the immanent world of contingency. To marvel at the existence of the world, or at the existence of language, is to imply the possibility of the nonexistence of the world or language. To wonder at the existence of the world is to place the "fact" of the existence of the world alongside all the other contingent "facts" of the world. To wonder at the existence of the world is to attempt to place oneself outside of the world, but this attempt can only occur as a conceptualization within the world and therefore becomes part of it. And even if the expression of the impossibility of ethics is said to point beyond the realm of possibility to the realm of necessity, both the expression itself and the sense that it shows something beyond itself are simply two of the many facts of the inescapable world.

Attempting to imagine ethics as absolute value is as noble and as futile as attempting to escape language by means of language. It can't be done, but it can be shown that it can't be done, and the pattern of its impossibility is said to give a glimpse of a world outside of language. Expressions of absolute values, absolute good, and the ultimate meaning of life are nonsense, but, for Wittgenstein, the *function* of nonsense is to point away from the relative world in which sense is made.

> I see now that these nonsensical expressions were not nonsensical because I had not yet found the correct expressions, but that their nonsensicality was their very essence. For all I wanted to do with them was just *to go beyond* the world and that is to say beyond significant language. My whole tendency and I believe the tendency of all men who ever tried to write or talk Ethics or Religion was to run against the boundaries of language. ("Ethics" 11–12)

Yet, no matter how successfully nonsensical language may *show* the possibility of a realm beyond sense, it can never cross over into that realm. Even nonsense in language is forever doomed to make sense of itself. Therefore, the attempt to escape the boundaries of our language, which are the boundaries of our world, is—ironically—"perfectly, absolutely hopeless" ("Ethics" 12).

Does this mean, then, that ethics is its own impossibility? The transcendental realm of ethics is defined by terms like "perfect" and "absolute," terms that have no meaning in the contingent world of "relative" values, therefore the transcendental project of ethics is marked by a "perfect" and "absolute" hopelessness. If it is to remain true to its own transcendence, ethics, it seems, must maintain the necessity of its own impossibility. In fact, ethics is identified *as* the necessity of its own impossibility. The figure traced is quite paradoxical. Wittgenstein starts with a basic distinction—call it immanence/transcendence, inside/outside, relative/absolute, contingency/necessity, sense/nonsense—and attempts to think the possibility of crossing over from the left side of this distinction to the right. The world of language in which this attempt is made is radically immanent. It is a world in which sense is made, in which every proposition implies its own negation, i.e., the possibility of its own nonexistence. The attempt to think ethics (defined as absolute value) is an attempt, made from within the contingent world of sense-making, to transcend the contingent world of sense-making. The inside stretches to become its own outside in order to see itself and know itself as absolute necessity. But the task is "hopeless," necessarily doomed to failure. The act of making sense is the act of making distinctions. The attempt to overcome the making of distinctions by making a distinction between making distinctions and not making distinctions is quite obviously impossible. In fact, it has the unintended consequence of expanding the boundaries and increasing the territory of the world of distinctions. For every inside that succeeds in seeing itself from its own outside, there is a further outside that can be discerned, distinguished, and designated. The inside turned outside is recaptured as an inside.

The transcendental/immanent distinction, coupled with the impossibility of escaping the domain from which this distinction is made, results in a vast and oppressive immanence, the inescapability of which is guaranteed by the attributes given to that side of the distinction that cannot be reached. The world thus becomes *absolutely* contingent. It cannot be otherwise than the fact that the world can be otherwise. The impossibility of necessity is necessarily the case. Transcendence guarantees the conditions for the possibility of immanence by removing itself from the field of observation, for if observed, it would disappear into the vast immanence it calls forth. And so immanence becomes the closed system of the world whose contingency is not contingent. We are left with a systemic solipsism. The outside is acknowledged as the absolute condition for the existence of the inside, but it remains supremely unknowable. It is the silence that delimits the world. "My work," Wittgenstein writes of the *Tractatus*,

> consists of two parts: the one presented here plus all that I have *not* written. And it is precisely this second part that is the important one. My book draws limits to the sphere of the ethical from the inside as it were, and I am convinced that this is the ONLY *rigorous* way of drawing those limits. In short, I believe that where *many* others today are just *gassing*, I have managed in my book to put everything firmly into place by being silent about it. (Letter to Ludwig von Ficker, cited in McGuiness 288)

This, then, is the final paradox. Since the search for ethics is described as the quixotic attempt to run up against the limits of language (Waismann

13) with no hope of ever occupying the position of the extra-linguistic, transcendental observer, there is nothing to be done but to resign oneself to the position one does occupy. The supreme power of transcendence, then, is its undoing. One cannot evoke the outside and demand a radical change of the world, because the only change that could satisfy the claims of absolute ethics would be the absolute destruction of the contingency that is the world. The result, in the words of Wittgenstein's friend Paul Engelmann, is "an ethical totalitarianism in all questions, a single-minded and painful preservation of the purity of the uncompromising demands of ethics" (109). But these demands, Wittgenstein makes clear, can neither be taught (Waismann 16) nor articulated in such a way as to have "the coercive power of an absolute judge" (Wittgenstein, "Ethics," 7). They are felt as the pressure simultaneously to accept and to distance oneself from the world as it is. To experience "the discrepancy between the world as it is and as it ought to be" (Engelmann 74) is not a mandate to rail against the world, but to realize "that that discrepancy is not the fault of life as it is, but of myself as I am" (Engelmann 76). For Wittgenstein, the ethical call is transcendental and absolute, but, ironically, its absoluteness is also its impotence. It does not result in moral precepts or political programs, it merely demands recognition of a simple fact: The world is as it is.

II

The world is as it is. If the givenness and valuelessness of the world leads to acceptance of the world as it is, then, as Wittgenstein recognizes, not only does the discrepancy between "I" and the world disappear, but so does the opposition between solipsism and realism (*Tractatus* 5.64). "The world is as it is" becomes identical with "the world is as I see it." The subject is neither a part of the world, nor does it stand opposed to the world, but rather is its horizon, beyond which the world ceases. "[T]he world is my world," as Wittgenstein says (*Tractatus* 5.641). For most, the simple equation I = World is an intolerable tautology, an eternal reproduction of the same and exclusion of the other. The "Not-I," the other as excluded middle, is squeezed out of the system and condemned to the unknowable realm that surrounds the world and allows for the definition of the limits of the world. If, however, ethics is proclaimed to be transcendental, then the domain of the ethical is simultaneously the domain of the other, and the question of the relationship of this ethical outside to the immanent inside becomes a meditation on the relationship of self to other. If the excluded other becomes invested in this way with ethical force, does it then take on messianic qualities and arrive as a thunderclap, as the static and noise come from the outside to disrupt the smooth functioning of the self-reproducing system? If so, how is one to heed this ethical call from the infinitely other? Can the Messiah be perceived as the Messiah, or is thunder just thunder? That is, how can one accommodate disruption and hear noise as something other than noise, without denying the alterity of the other? How can the transcendental call of the other be heard without reproducing the dialectic of inclusion and exclusion that forever pushes the other back outside the system and domesticates transcendent ethics in terms of moral codes or political prescriptions?

One could construe this situation, with Jacques Derrida, as the "original tragedy":

> *My world* is the opening in which all experience occurs, including, as the experience par excellence, that which is transcendence toward the Other as such. Nothing can appear outside the appurtenance to "my world" for an "I am."

Or pose it as the essential philosophical question:

> [W]hy is the essential, irreducible, absolutely general and unconditioned form of experience as a venturing forth toward the other still egoity? *Why* is an experience which would not be lived as *my own* (for an ego in general, in the eidetic-transcendental sense of these words) impossible and unthinkable?

But one "cannot answer such a question by essence, for every answer can be made only in language, and language is opened by the question" (131). The condition—the tragedy, if you will—of language is the originary violence that both allows us to distinguish ourselves from others and condemns us never to cross the limits of that distinction. "Space" is the "wound and finitude of birth . . . without which one could not even open language" (112). Language can never be "weaned" from "exteriority and interiority," one could "never come across a language without the rupture of space," since "the meanings which radiate from Inside-Outside, from Light-Night, etc., do not only inhabit the proscribed words; they are embedded, in person or vicariously, at the very heart of conceptuality itself" (113). Therefore, any attempts to overcome language and violence lead only to their replication. "Discourse . . . can only *do itself violence,* can only negate itself in order to affirm itself, make war upon the war which institutes it without ever *being able* to reappropriate this negativity, to the extent that it is discourse" (130). The first word is the first wound in a chain of wounds that never heal. "Violence appears with *articulation*" (147–48).

Nevertheless, precisely because of the inescapability of violence and of language, Derrida resists in Levinas what we have seen in Wittgenstein, the absolute transcendence of the other. As we saw in Wittgenstein's acceptance of the world, far from guaranteeing the otherness of the other, positing absolute transcendence only collapses it into the world of absolute immanence. For Derrida, what *does* guarantee the otherness of the other is its ability to reappear in the world distinct from the world. The inside/outside distinction that arises with language and constitutes the system and its other cannot be obliterated in or by language, but it can be replicated and can reenter the system, the inside, as "trace" of the outside in the form of an ego/alter ego distinction. Because "I" can only perceive "my" world, the other has to be presented in it for me to be exposed to it, and only by perceiving the other, analogically, as an other ego, can "I" perceive it *in* "my" world as something *other* than my world. "If the other were not recognized as a transcendental alter *ego*, it would be entirely in the world and not, as ego, the origin of the world. . . . If the other was not recognized as ego, its entire alterity would collapse" (125).

There is a tension at this point in Derrida's reading of Husserl, Levinas, and Heidegger that might lead one into the temptation of thinking redemption. On the one hand, with the original tragedy (or should we say *sin*) of language comes "irreducible violence" (128) and universal "war" (129), which, as we have seen, only reproduces itself, even when it attempts the opposite.[3] Yet, "this necessity of speaking of the other as other, or to the other as other" is not only "the transcendental origin of an irreducible violence," it is "at the same time nonviolence, since it opens the relation to the other" (128–29). The initial word inevitably institutes violent separation, but that originary alienation is required if there is to be the possibility of a nonviolent relation between ego and other as alter ego.

One can catch a glimpse here of a familiar teleology. History may be the "infinite passage through violence" (130), but within history, that is, "[b]etween original tragedy and messianic triumph there is *philosophy,* in which violence is returned against violence within knowledge, in which original finitude appears, and in which the other is respected within, and by, the same" (131). Is this respect that is due the other nonviolent, and if so, how are we to think nonviolence within a history defined *as* violence? Surely philosophy does not transcend history and language. Within history, the messianic triumph can never occur, not even as philosophy. But is Derrida saying that even if the Messiah never arrives, even if we are to believe in the impossibility of his arrival, are we to figure the ego and the alter ego as the lion and the lamb in order to hold the image of reconciliation as pledge and hope? Is this pledge capable of being redeemed? As respect? Is Derrida's closing soliloquy on Heideggerian Being at the end of this essay to be read as a prayer to the unattainable but eternally longed-for Unity that lies beyond the distinction between distinction and unity?

These are temptations that Derrida is normally said to resist. More to the point, these are temptations that Derrida normally observes and condemns in others (alter egos?). But redemption does have its allure, and it is out of this glimmer of a utopian moment in Derrida that Drucilla Cornell attempts to construct a "quasi-transcendental" ethics in her study, *The Philosophy of the Limit*:

> For my purposes, "morality" designates any attempt to spell out how one *determines* a "right way to behave," behavioral norms which, once determined, can be translated into a system of rules. The ethical relation, a term which I contrast with morality, focuses instead on the kind of person one must become in order to develop a nonviolative relationship to the Other. The concern of the ethical relation, in other words, is a way of being in the world that spans divergent value systems and allows us to criticize the repressive aspects of competing moral systems. (13)

Here it is clear that Cornell wants to establish a hierarchical relationship between ethics and morality. Morality, which is to be subordinated to ethics, is equated with the enunciation of behavioral norms and the generation of a system of rules. The ethical relation, on the other hand, does not manifest itself as discourse. Rather, it is embodied by a carefully semi-specified way of being that allows it to sit in judgment on moral systems. Over and against morality—which, as the enunciation of rules and norms, inextricably forms part of any system—the ethical relation would seem to occupy a transcendental position, the simple "outside" of any articulated moral code. However, the rudimentary specification of this ethical relation (e.g., the use of two strategic adjectives) reveals that what surveys "divergent value systems"—ethics—is itself rooted in a value system. The ethical relation dictates that we have a "nonviolative" relationship to the other, and the way of being that results from this relationship specifies that we are to criticize "repressive" aspects of the moral systems we observe. Both adjectives suggest a moral code: "Thou shalt not violate nor repress the Other." The "outside" of morality now finds itself "inside" a specific, if only partially specified, moral system. This passage, then, traces the figure of a dual transcendence. The ethical relation "bootstraps" its way out of its underlying code to serve as the quasi-transcendental perspective from which moral systems can be judged. The ethical relation cannot, however, judge the code with which it is implicitly linked—i.e., it cannot criticize nonviolence—without ceasing to exist as the quasi-transcendental self-reflection of morality. The ethical relation, then, displays itself as both the master and slave of morality.

One is tempted to ask why the relation to the other has to be nonviolative. Why does nonviolence (a term she uses interchangeably with non-violative) serve as the unquestioned ground for an ethics that in turn is to serve as the ground for political (or, at least, juridical) action? The answer lies in Cornell's fear for the fate of the other in the inexorable and impersonal grindings of legal machinery. As a legal theorist, she is concerned with establishing a distinction between justice and existing law, since in her view, social criticism hinges on the deconstructibility of the actual existing law, and the deconstructibility of law depends on the undeconstructibility of justice.[4] Cornell argues that for the legal positivism of a Stanley Fish, no such distinction, no "true *difference* from the system," exists. "Because for Fish there is no divide between justice and law, the deconstruction of law is not possible," making, therefore, "social criticism and radical transformation" also impossible (145). The tautology that results from the equation "justice = law" needs, according to Cornell, to be disrupted so that justice may be dislodged from its all-too-close proximity to the law. But precisely because absolute transcendence, absolute distance between the system and its outside, collapses into effective immanence (as we saw happen with Wittgenstein), Cornell's transcendental other must have a way of entering into the system while still retaining its status as outsider.

The dilemma Wittgenstein faced was the impossibility of occupying the transcendental position dictated by ethics, while still remaining in the world that is to be judged by this position. That is to say, he was confronted with the logical paradox of attempting to be simultaneously outside and inside the system, a paradox he could only resolve by embracing immanence in the name of transcendence. Cornell feels that for a transcendental perspective to be ethically and politically effective, it cannot simply remain as a godlike position supremely distant from and outside of the system. The solution to the paradox can neither be apocalypse nor resignation. She therefore reproduces the system/environment (inside/outside) distinction *within* the system itself and anthropomorphizes it as an ego/other relationship, in which it can be said that the *absence* of the other within the system occupies a form of transcendence. For a system to define itself, it must distinguish itself from that which it is not. What is thereby excluded from the system is defined as the system's other. But, says Cornell, there is a "responsibility to memory" involved in this act of self-definition by way of exclusion. The "system, through the critical observer, is called to remember its own exclusions" (149). The "trace" of the other is the history of the other's exclusion. The act of remembrance—or mourning, if you will—becomes, then, the quasi-transcendental position that is accessible and capable of being occupied within the system, and from which the system can be seen as the other of its other.

The argument is compelling when anthropomorphized as ego/alter in this way. The other as transcendental perspective lingers like a bad conscience in the shadows of the world from which it has been banned. Thus victimized by the machinery of exclusion, it no longer is thought of as the silence that surrounds the world, but as the *silenced* one *in* the world. And as the silenced, oppressed, and marginalized, not only can the other serve as the extramundane critique of the world as it is, but by way of empathy and affiliation, the quasi-transcendental position of the other can be imaginatively occupied by those *not* excluded from the system, allowing the system to critique itself in the name of justice. In this way, the aporia of a Wittgensteinian (or Levinasian) transcendental notion of ethics is overcome.

But the question remains, who or what is the other? More to the point, if the other is defined as the silenced and the excluded, who or what *names* the other if not the very system that silences and excludes? Cornell is of course aware of the inherently paradoxical nature of attempting to name the other, or attempting to hear the other name itself, and she realizes that Derrida, whom she takes as her guide in this matter, "leaves us with the paradox that the Saying can never be said"; yet she still believes that her attempt to "nam[e] the ethical force of the philosophy of the limit" remains "true to the paradox" (89, 90). But can it remain true to the paradox? If violence is original sin, if the world of selves and others, systems and environments, arises out of the violence of distinguishing and naming, then the attempt to base an ethics on the nonviolent relationship of self and other represents a desire for the healing of the wounds of existence—in short, for redemption. Now Cornell realizes that redemption does not remain true to the paradox; it obliterates it. It can only be present as the promise of an indefinitely postponed (first or second) coming, for if the Messiah, wearing the robes of Justice, were actually to arrive, the world of desire would be destroyed. Justice can only be Justice Deferred. There can be no paradise on earth, because paradise annihilates the earth that desires paradise as surely as a book on ethics would annihilate all the books that desired ethics. And yet, Cornell's desire to name the ethical force is founded on the desire for what she calls a "radical transformation" of the system (142, 145). What needs to be questioned, then, is whether this desire for radical transformation is simply the wish to give direction to necessary change, or whether it does not harbor a certain longing for that final unchanging change, that prescription against change, that comes with any and all Messiahs.

The problems with Cornell's raising of the other to quasi-transcendental heights become apparent, I believe, in her discussion of *Roe v. Wade*. Here, the position of the other is said to be occupied by woman, and the "critical observer" is called upon "to remember the history in which women did not have the right to an abortion" (149). Justice Blackmun (as author of *Roe v. Wade*) is praised for "imaginatively recollect[ing] a legal norm from within our heritage that would allow us to make crucial distinctions about the status of the fetus for the purposes of law," and Justice Rehnquist *(Webster v. Reproductive Health Services)* is chastised for "substitut[ing] his own standards in lieu of those which already existed" (150). In particular, Rehnquist, in his advocacy of the rights of the fetus over the rights of women, is accused by Cornell of deliberately disregarding "the genealogical considerations demanded by integrity. These considerations are demanded by the call of the Other for Justice" (152). As a defense of abortion rights, this seems to be a dangerous tactic.[5] Could it not be claimed by antiabortion activists that Rehnquist did in fact heed the "call of the Other for Justice," that at least at the time of *Webster v. Reproductive Health Services*, the "Other" *was* the fetus? Once the other is identified as woman and the quasi-transcendental position is called upon to acknowledge "women's demand for the right to abortion," has not the other—Woman—become the quasi-transcendental subject, if not the plain old transcendental subject of history (since, as empirical subjects, women are quite divided on the issue), and could it not be argued, using Cornell's own reasoning, that the responsibility to memory now urges us to recognize the (literally) excluded fetus as the new other, the new quasi-transcendental position from which the critical observer is to utter judgments? Is there not a dialectic of self and other that prohibits ultimate calls to Justice based on the other, as if the other were always the same?

The issue ultimately rests on what we think the nature of change is. Does the call to change a particular system presuppose ultimate and inevitable contingency (i.e., things can always be other, even after they have been changed), or is change, when intensified into radical transformation, conceived of as arriving at a final destination? Is the imperative for change, in other words, another term for contingency, or is it contingency's termination? With regard to abortion, what would a radical transformation be? What resolution to the conflict between those who claim priority for the rights of women and those who claim priority for the rights of the fetus can there be that is not a violent repression of a perceived right? We have here a paradigmatic instance of what Lyotard calls a "differend,"[6] and no adjudication of the matter can avoid violence. What purpose does it serve, then, to camouflage this fact with a quasi-transcendental construction of justice? Cornell invokes justice as the absent judge. By being invoked, justice is inhabited, and the voice of the other is incarnated in memory and ventriloquized; but is such a ventriloquy ever the voice of the other, and is the invocation of justice (or Justice) ever anything more than the enunciation of the law? Once a law is enunciated, another other has been prepared for future ventriloquies, and the cycle of violence that is history—no, that is politics—is continued. If guaranteeing abortion rights is a necessarily violent, political act, why "dress it up" in moral discourse? Is not this desire for an ethics that invokes the absolute authority of the outside in its attempt to banish violence really a desire for the end of the "unclean" world of disputation and politics? And if so, is this not ultimately a paradoxical position—one is tempted to say a conservative position—for Cornell to espouse, a position that harbors a traditional fear of the loss of transcendental authority even as it attempts to critique traditional transcendental arguments?

In this attempt to figure the outside as the returning other, we can see a concern with opening an otherwise closed system to the possibility of change. The assumptions that seem to guide Cornell's ethical imperative are 1) that closure precludes change because 2) change must be morally guided. On this view, closed systems naturally tend toward equilibrium, and in order for change to occur, the (correctly imagined other as) environment must *instruct* the system, must determine change, not simply act as its occasion. The other—at least as ventriloquized by Cornell— authorizes Blackmun, not Rehnquist, to be its spokesperson. But such an authorization would seem to subject the legal system to moral oversight, a position the system has fought long and hard to escape. The "ethical moment" threatens to become, then, the moment that jeopardizes the autonomous self-reproduction of the system by dissolving the clear distinction between system and environment. The law ceases to be the law when the ethically occupied other lays it down. It becomes a commandment.

III

As is well known, in his "Politics as a Vocation," Max Weber distinguishes between an ethics of responsibility and one of ultimate ends, adding that political action can never be successfully linked to the latter. "He who seeks the salvation of the soul, of his own and of others," he writes, "should not seek it along the avenue of politics, for the quite different tasks of politics can only be solved by violence" (126). This disjuncture between the spiritual and the material, the religious and the secular, is a result of the rationalization and functional differentiation of modernity. A "specialization of ethics" follows from the fact that "[w]e are placed into various life-spheres, each of which is governed by different laws" (123). In such a modern world, the dreamed-of universal ethics or moral integration of society (Habermas) seems to be gone forever. Whereas most, including Weber himself, have seen this compartmentalization of morality as a cause for anxiety, Luhmann unabashedly endorses it. For him, it is the only way to preserve what he considers to be the hard fought and improbable victory of systemic autonomy that marks differentiated modernity. Unlike the traditions of early German sociology and Western Marxism, which describe modernity (with varying degrees of nostalgia) in terms of reification, rationalization, and colonization of prerational lifeworlds, Luhmann assesses modern differentiation positively, without necessarily adhering to 1950s fantasies about the inevitable benefits of the process of modernization in individual function systems. In his view, modern society is organized as a horizontally (not hierarchically) ordered plurality of autonomous social systems, with no one system able to control or dictate to any other system how it is to discharge its function. By way of this functional differentiation—the relentless self-division of society into the specialized function-systems of politics, economics, art, science, law, religion, pedagogy, and so forth—modernity develops the capacity to deal with increased environmental complexity (which it also helps create) through the organization of "resilient"[7] formal relationships among basic elements, i.e., a formal organization with the flexibility and capacity for change that allows it to withstand environmental assaults. Recall that Luhmann, adapting Maturana's notion of autopoietic closure, thinks of social systems as operationally closed with respect to information. A system runs blind, so to speak. It does not receive informational inputs from its environment; rather, environmental "perturbations" simply serve as catalysts for the operations of a system's internal organization. The outside "impinges" on a system, but remains unknown, unoccupied, unthought. The only position the system *can* occupy is the position of the system. Luhmann, therefore, is not interested in investing this outside with moral agency, nor is he concerned with constructions of high culture (Adorno) or lifeworlds (Habermas) that could somehow serve within society as society's other. His concern, on the contrary, is with the continued self-reproduction of modernity's differentiated function-systems.

Since Luhmann considers communication—not individuals, not subjects, not humans—to be the basic element of social systems,[8] the notion of systemic closure and functional differentiation can be conveyed by saying that the "language" of one system cannot be adequately translated into the "language" of another system. Much like Wittgenstein's language games or Lyotard's genres, they are incommensurable, a fact that guarantees their autonomy (or, as Lyotard would say, a fact that guarantees the lack of a grand, totalizing narrative). A system's communication is channeled and directed (i.e., complexity is reduced and managed) by its unique, binary code.[9] Each system uses environmental perturbations as the "excuse" to generate information by way of its own code, and these codes, as Luhmann puts it, stand in an "orthogonal" relation to one another. Simply put, they do not overlap. Science may process information according to a true/false schema, art according to a beautiful/ugly or interesting/uninteresting one, and economics according to a profitable/unprofitable one, but this does not make what is true automatically both beautiful and profitable, or what is profitable both beautiful and true. We are not dealing with homologies.

The insistence on the incommensurability, autonomy, and autopoietic closure of social systems like science, economics, and politics—which should not imply a lack of interaction or interpenetration, but rather a means of establishing identities—is of crucial importance for Luhmann's handling of the problem of morality in modern society. In fact, in his various treatments of the topic,[10] Luhmann's political commitments clearly emerge, though, characteristically for him, they emerge in the form of a circle. What presents itself, in Luhmann, as descriptive of modernity also takes on the force of a prescriptive. The description of modernity as differentiated needs to be read both as an empirical fact—"differentiation exists"—and as an imperative—"differentiation ought to (continue to) exist." That differentiation exists and ought to exist translates, then, into a political injunction: "Thou shalt not de-differentiate!"[11] This perceived imperative, then, dictates Luhmann's concern with ascribing limits to the applicability of the moral code.

This problematic relationship between descriptions and prescriptions is reflected in the way binary distinctions both create and "unfold" paradoxes. In Luhmann's view, morality also operates in society as a communication steered by a binary code—good/bad, articulated in terms of approval or disapproval. But whereas the other codes Luhmann analyzes (government/opposition, profitable/unprofitable, true/false) are housed, so to speak, in the institutional structures of social systems (the state, the capitalist marketplace, the university and academic publishing industry), and whereas the legal system has taken over the function of determining social norms (along the axis legal/illegal), the moral code has detached itself from its premodern locus in religion and has become a self-replicating, parasitic invader of the various modern, functionally differentiated, social systems.[12] Luhmann actually writes of morality as a bacterial infection, and concedes that "like bacteria in bodies, morality can also play a role in function-systems." But if it is not to destroy the system it inhabits, it must orient itself toward "the structural conditions of the respective function-systems" and not according to some "metacode" that aims at totalization ("Ethik" 431). The danger comes, according to Luhmann, when the moral code—good/bad—attaches itself "isomorphically," one might say, to the prevailing codes of the respective function-systems, when it seeks, that is, to impose a binding translation of "true" or "government" or "profitable" into "good" (or "bad"). Such a debilitating, moral "infection," or parasitic overlay of the good/bad grid, would paralyze the autonomous functioning of the system, eventually causing it to lose its identity and disappear.

The ever present temptation to moralize politics is an example of the danger of such an infection. If, as Luhmann contends, the political system in modern parliamentary democracies orients itself toward the distinction government/opposition, then it cannot allow the value "good" to attach itself only to the governing party (or opposition party) and still exist as an autonomous social system. "Neither the government nor the opposition," he writes:

> should entangle the model of government/opposition in a moral scheme in the sense that one side (our's) is the only good and respectable one, while the other side acts immorally and reprehensibly. For this would inhibit the very idea of a change from government to opposition as such and the idea that democratic rules work. ("Future" 237–38)

Thus, the incongruence of codes formally guarantees the circulation of power and thereby the legitimacy of political decision making in the political system. In a similar manner, the value "good" cannot be allowed to be coterminous with "legal," for otherwise how could one challenge existing law without exposing oneself to moral condemnation, or, what amounts to the same thing, how could one continue the replication of the legal system through legal (not moral) communication? If one does not preserve the "orthogonal" relation between the moral and legal codes, how else—to return to the abortion debate—could one distinguish between the morality and legality of abortion, a distinction then-candidate Bill Clinton made in his acceptance speech at the 1992 Democratic National Convention when he affirmed his opposition to abortion while still defending the right of a woman to choose. Given the climate that surrounds this debate, such a statement by a politician strikes most as hypocritical, dishonest, and opportunistic—and maybe from the perspective of "sincerity" or "authenticity," whatever those qualities may be, it was. But from the perspective of the political system in which it was uttered, it could also be construed as distinctly "modern" in the sense of a radical—and radically desired—disjuncture between legal and moral codes.

If, as I have claimed, there is a prescriptive element in Luhmann's treatment of morality, and if, as Luhmann claims, the moral bacterium can have positive effects, then the question arises: To what *should* the moral value "good" attach itself, if not exclusively to one side or the other of a given code? It is clear that the value "good," in Luhmann's own description of modernity, should attach itself to the very *distinction* each code embodies, i.e., to the *difference* "government/opposition" (or "legal/illegal," and so forth) that defines the political system and by which the system communicates and reproduces itself. Preserving the autopoiesis of the system (not its historically contingent structure)[13] by preserving the independent functioning of its code becomes Luhmann's moral imperative, an imperative he expresses with yet another distinction, the one between morality and an ethics that is set up to serve as morality's self-reflective conscience.

This distinction between morality and ethics is historically conditioned. Ethics as the reflection theory of morality becomes necessary when caste-based moral codes of conduct, defined by Luhmann as the unity of morality and manners, gives way, along with the stratified ("feudal") social organization in which it flourished, to increasingly complex, differentiated modernity ("Ethik" 416). In both its utilitarian (Bentham) and Kantian varieties, this new emphasis on ethical reflection is registered as the necessity for establishing criteria for choice ("Ethik" 413). In other words, ethics becomes formalized, moved from a consideration of the moral "fiber" or substance of an individual to a consideration of action in the face of competing alternatives. The Kantian solution, as is well known, relies on the validity of the transcendental/empirical distinction and therefore is, in Luhmann's view, no longer tenable. Historically speaking, the circle that this distinction interrupted is back in operation because of the demise of the transcendental subject. Any renewed efforts to determine the function of morality by utilizing the morality/ethics distinction must locate ethics—i.e., ethically determined choice—inside the system called society. Ethics, then, as a decision-making process, is not anchored in free-floating, morally educated subjectivity, but in the need of social systems to protect themselves from the effects of morality ("Ethik" 371).

The effects Luhmann fears can be elucidated historically by listing the countless crusades, wars, inquisitions, and persecutions that moral discourse has fueled. By acting as mediator between morality and society, ethics is charged with minimizing the devastation morality is capable of unleashing. This attempt to shield society from the consequences of moral communication explains, in part, the limited and formal definition of freedom and democracy with which Luhmann operates. As has been noted, social systems exist and reproduce themselves by virtue of communication. Communication is defined not as the transfer of information from active producer to passive receiver, but as the production of information through choice on the part of the recipient. Communication could be said to be the result of continuously constructing the distinction between information and noise. Communication offers itself then as connectivity (Anschlußfähigkeit), as the opportunity to continue or discontinue communication. Freedom arises in systems as the ability to affirm or reject communication, and democracy is thus defined as the precariously evolved formal structure of differentiation that holds open the possibilities for affirmation and rejection, assent or dissent, in the political system. The resulting democratic principle par excellence is defined as maximalization of choice. Modernity, as the functional differentiation of social systems that reproduce themselves by way of communication, represents a highly evolved structure that continuously enforces the necessity for selection, and therefore continuously reproduces freedom in this highly abstract and formal sense.

Now, morality, too, is a form of communication, and therefore morality engages in the production of the choice between affirmation and rejection as well. But the moral code, Luhmann contends, has the additional function of inhibiting or "suggesting away" ("*wegzusuggerieren*") the freedom it produces by coding approval and disapproval of the consequences of communication. Its aim, in short, is to eliminate choice (even as it produces it) by preselecting affirmation and rejection ("Ethik" 439–40; "Paradigm Lost" 91). That is, morality, with its code of approval/ disapproval, attempts to limit the choice it cannot help but automatically engender. It attempts to impose its means of reducing complexity on the systems it inhabits, i.e., it attempts to replace a "legitimate," system-specific means of generating and processing information with an "illegitimate," totalizing and parasitic one. This description of morality is, of course, not a self-description of morality, but rather comes from the perspective of a social system's general theory of social systems. Its interest is the preservation of the autopoiesis of the system it describes (including, it must be added again, the autopoiesis of modernity as functional differentiation). *Ethics*, therefore, described from this systemic perspective, is seen as a kind of *immune system* or *on/off switch*, and we are advised that "perhaps the most pressing task of ethics is to warn against morality" ("Paradigm Lost" 90). But, as a version of the traditional intolerance of intolerance, it can only do so as paradox. As a reflection of and on morality, ethics operates with morality's code, only what it subjects to this code is morality itself. Because of morality's limitation of freedom, *its* freedom must be limited, or, as Luhmann puts it, because of its negative, violent effects, morality undergoes a (violent) civilizing process ("Ethik" 435, 436). Thus, by way of ethics, morality is called upon to discipline itself for the sake of the system. "No progress without paradox," Luhmann notes ("Paradigm Lost" 91).

If we remain within the immanence of systems that Luhmann not only advocates, but sees as inescapable, we are left with this paradox. Ethics emerges as the by-product of a system's attempt to preserve its own reproduction from the ravagings of moral infection. The only moral preselection said to be ethically permissible is the preselection that guarantees the freedom of selection. Thus ethics, though it can make moral judgments regarding morality, reaches its limit when it attempts to judge itself, and it must resign itself to this inherent limitation. As a reflection theory of morality, not society, it must use (a particular description of) society as its ground, shielding that ground from internal and external threats. In this way, Luhmann resists transcendental temptations—or at least attempts to. His minimalist liberalism—intent on preserving operations, not contents—and his formal definitions of freedom and democracy leave no room for talk of emancipation as an achieved, or even as a utopically desired state. Democracy, according to Luhmann, is not "about emancipation from societally conditioned tutelage, about hunger and need, about political, racist, sexist and religious suppression, about peace and about worldly happiness of any kind" ("Future" 231); it is simply the prerequisite for the political, economic, and legal observation and discussion of such problems. Democratic discussion cannot successfully mandate outcomes, nor does it proceed along the lines of consensus or predicated rules, other than the "rules" of the various observing systems that are and will remain at odds with each other. Democratic discussion simply reproduces the conditions for its own possibility, and the rights we fight for, it would seem, arise, when they arise, as the by-products of this continuous activity. It is not a utopic vista that Luhmann paints, simply an improbable and, he seems to think, highly fragile condition. And it is *this* condition—not project—of modernity that he invests with an ethical imperative.

Notes

1. For Maturana's own account of the genesis of the concept of autopoiesis, see his introduction to Maturana and Varela, *Autopoiesis and Cognition,* xi–xxx. See also the interview, "Gespräch mit Humberto R. Maturana," in *Zur Biologie der Kognition,* ed. Volker Riegas and Christian Vetter. A description of the color-perception experiments can be found in Maturana and Varela, *The Tree of Knowledge,* 18–23. For a critical assessment of Maturana's use of the empirical evidence, see Riegas.

2. See Luhmann, *Social Systems,* for a full elaboration of his social theory; see also Luhmann, "The Autopoiesis of Social Systems" in *Essays on Self-Reference.* For introductions to Luhmann in English, see Dietrich Schwanitz, "Systems Theory According to Niklas Luhmann: Its Environment and Conceptual Strategies" (in this volume); Harro Müller, "Luhmann's Systems Theory as Theory of Modernity"; and Eva Knodt's forthcoming introduction to the English translation of *Soziale Systeme.* For Maturana's critique of Luhmann's adaptation of his concept, see Riegas and Vetter, "Gespräch," 39–41.

3. "The very elocution of nonviolent metaphysics is its first disavowal" (148).

4. Cornell refers here to Derrida's essay, "Force of Law: The 'Mystical Foundation of Authority,'" in *Deconstruction and the Possibility of Justice,* ed. Drucilla Cornell, Michel Rosenfeld, and David Gray Carlson.

5. Legally, the crisis Cornell identifies seems to have eased, at least at the level of the Supreme Court. With the two Clinton appointments, *stare decisis* seems momentarily adequate as a defense of *Roe v. Wade.* Pragmatically and politically, the battle needs to be waged *against* moral discourse, not for it. The task is to show how

moral arguments are used as weapons against women's (and gays') legitimate (i.e., precedented) demands for civil rights. Once arguments about the "other" enter the picture, then the "silenced Christian majority" as the other of "liberal, permissive society" is conveniently handed a weapon with which it can pummel us.

6. "A case of differend between two parties takes place when the 'regulation' of the conflict that opposes them is done in the idiom of one of the parties while the wrong suffered by the other is not signified in that idiom" (Lyotard 9).

7. Operational closure need not be synonomous with homeostasis. On the contrary, the *resilience* of a system, to use C. S. Holling's teminology, depends on its ability to "absorb" and accommodate itself to environmental disturbances, to "keep options open," to "emphasize heterogeneity," and not on its ability to take instructions (21). Only by formally rigorous and self-referentially operative closure can the system (indirectly and unpredictably) accommodate environmental perturbations, including the perturbations of an ethical call.

8. Again, for a detailed description, see *Social Systems*, 137–75. For a brief discussion in English, see *Ecological Communication*, 28–31, but see also *Essays on Self-Reference*, 80–106.

9. On binary codes, see *Ecological Communication*, 36–50.

10. On morality, see "Ethik als Reflexionstheorie der Moral." In English, see "Paradigm Lost"; "Politicians, Honesty, and the Higher Amorality of Politics"; and *Ecological Communication*, 127–32.

11. Cornell's appreciation and critique of Luhmann centers on his optimistic and hopeful assessment of functional differentiation. In her view, remnants of stratification can be accounted for by viewing gender hierarchy as a system, a system that should be destroyed. See her "The Philosophy of the Limit: Systems Theory and Feminist Legal Reform" in *Deconstruction*, ed. Cornell, Rosenfeld, and Carlson, 68–91. Luhmann would tend to see the continued existence of gender hierarchy in the same way he sees morality, as a threatening and parasitic invader of functional systems (see discussion of morality below).

12. On morality's "free-floating" status, see "Ethik," 421, 434–35.

13. Though it seems clear that the imperative to preserve any particular system's code is the "micro" equivalent of the "macro" imperative to preserve functional differentiation, i.e., modernity, as such.

Works Cited

Cornell, Drucilla. *The Philosophy of the Limit*. New York: Routledge, 1992.

Cornell, Drucilla, Michel Rosenfeld, and David Gray Carlson, eds.
    Deconstruction and the Possibility of Justice. New York: Routledge, 1992.

Derrida, Jacques. "Violence and Metaphysics: An Essay on the Thought of

Emmanuel Levinas." *Writing and Difference.* Translated by Alan Bass. Chicago: University of Chicago Press, 1978.

Engelmann, Paul. *Letters from Ludwig Wittgenstein: With a Memoir.* Translated by L. Furtmüller. New York: Horizon, 1968.

Holling, C. S. "Resilience and Stability of Ecological Systems." *Annual Review of Ecology and Systematics* 4 (1973): 1–23.

Horkheimer, Max, and Theodor W. Adorno. *Dialectic of Enlightenment.* Translated by John Cumming. New York: Seabury Press, 1972.

Luhmann, Niklas. "The Cognitive Program of Constructivism and a Reality That Remains Unknown." In *Selforganization: Portrait of a Scientific Revolution,* edited by Wolfgang Krohn, Gunter Kuppers, and Helga Nowotny. Dordrecht: Kluwer Academic, 1990.

———. *Ecological Communication.* Translated by John Bednarz Jr. Chicago: University of Chicago Press, 1989.

———. *Essays on Self-Reference.* New York: Columbia University Press, 1990.

———. "Ethik als Reflexionstheorie der Moral." In *Gesellschaftsstruktur und Semantik: Studien zur Wissenssoziologie der modernen Gesellschaft.* Vol. 3. Frankfurt: Suhrkamp, 1980–94.

———. "The Future of Democracy." *Political Theory in the Welfare State.* Berlin and New York: Walter de Gruyter, 1990.

———. "Paradigm Lost: On the Ethical Reflection of Morality." *Thesis Eleven* 29 (1991): 82–94.

———. "Politicians, Honesty, and the Higher Amorality of Politics." *Theory, Culture and Society* 11, no. 2 (1994): 25–36.

———. *Social Systems.* Trans. John Bednarz Jr. with Dirk Baecker. Stanford: Stanford University Press, 1995.

Lyotard, Jean-François. *The Differend: Phrases in Dispute.* Translated by Georges Van Den Abbeele. Minneapolis: University of Minnesota Press, 1988.

McGuinness, Brian. *Wittgenstein: A Life. Young Ludwig, 1889–1921.* Berkeley: University of California Press, 1988.

Maturana, Humberto R. "The Biological Foundations of Self-Consciousness and the Physical Domain of Existence." In *Beobachter: Konvergence der Erkenntnistheorien?* edited by Niklas Luhmann, Humberto Maturana, Mikio Namiki, Volker Redder, and Francisco Varela. München: Wilhelm Fink, 1990.

Maturana, Humberto R., and Francisco J. Varela. *Autopoiesis and Cognition: The Realization of the Living.* In *Boston Studies in the Philosophy of Science* 42. Dordrecht: Riedel, 1980.

———. *The Tree of Knowledge: The Biological Roots of Human Understanding.* Boston: Shambhala, 1987.

Müller, Harro. "Luhmann's Systems Theory as Theory of Modernity." *New German Critique* 61 (1994): 39–54.

Riegas, Volker. "Das Nervensystem—offenes oder geschlossenes System?" In *Zur Biologie der Kognition,* edited by Volker Riegas and Christian Vetter. Frankfurt: Suhrkamp, 1990.

Riegas, Volker and Christian Vetter. "Gespräch mit Humberto R. Maturana." In *Zur Biologie der Kognition,* edited by Volker Riegas and Christian Vetter. Frankfurt: Suhrkamp, 1990.

———, eds. *Zur Biologie der Kognition: Ein Gespräch mit Humberto R. Maturana und Beiträge zur Diskussion seines Werkes.* Frankfurt: Suhrkamp, 1990.

Waismann, Friedrich. "Notes on Talks with Wittgenstein." *The Philosophical Review* 74 (1965): 12–16.

Weber, Max. "Politics as a Vocation." In *From Max Weber: Essays in Sociology,* edited by H. H. Geerth and C. Wright Mills. New York: Oxford, 1946.

Wittgenstein, Ludwig. "A Lecture on Ethics." *The Philosophical Review* 74 (1965): 3–12.

———. *Tractatus Logico-Philosophicus.* Translated by D. F. Pears and B. F. McGuinness. London: Routledge, 1961. Parenthetical references in the text are to propositions, not page numbers.

5. Rethinking the Beyond within the Real (Response to Rasch)

Drucilla Cornell

I want to thank William Rasch for providing me with the opportunity to respond to his thoughtful paper. I will proceed as follows: First, I will discuss the meaning I give to the word *ethical,* particularly as I use it to describe my own feminism as ethical feminism. Second, I will seek to clarify my own understanding of the "limit" as it challenges the traditional divide between immanence and transcendence. Third, I will return to my understanding of justice as the limit, specifically as this informs my analysis of Justice Blackmun's opinion in *Roe v. Wade.*[1]

My use of the word *ethical* in my definition of *ethical feminism* is indeed quite close to Luhmann's own understanding of ethics. For Luhmann, as Rasch rightly points out, ethics "emerges as the by-product of the system's attempt to preserve its own reproduction from the ravages of moral infection." Luhmann, of course, has a systems analysis of the rise of ethics. For Luhmann, ethics and modern moral systems both arise out of the end of stratified differentiation in which the question of what constitutes the good life could be answered by an appeal to established hierarchies and the characters associated with stations of life.[2] Paradoxically, the ethical serves as the warning against the systems of morality that produce it. Ethics, for Luhmann, is found in the need of social systems of modernity to protect themselves from the worst effects of moralizing, particularly as these inform our political battles.

I am deeply sympathetic to Luhmann's warning against the violence that has been unleashed by the so-called "civilizing process of morality." The paradox inherent in morality that I am calling the "ethical" and that Luhmann calls "ethics" is that morality's limitation of freedom demands that its freedom be limited. The classic example, of course, of the danger of moralizing in the political realm is best demonstrated in the political and purportedly legal battle about abortion. As we will see, Rasch misunderstands me when he argues that I am making moral arguments to defend the right of abortion. I am insisting instead that the question of abortion be decided legally, as a matter of the woman's right. But I will return to that discussion shortly. For now I want to stress that my own understanding of the ethical is close to Luhmann's, specifically as I warn against feminism's own mistaken conflation of the moral and the political. I also agree that the ethical is paradoxical and that it is tied to the demand of morality that it be disciplined for the sake of the system it tries to maintain.

Within feminism as a political movement, there is a specific danger associated with moralizing. Not only does the limitation of freedom imposed by morality impede the very process of unleashing the feminine imaginary that I make the heart of the matter of feminism, it also has the effect of dividing women among themselves. Specifically, the struggle to resymbolize the feminine within sexual difference beyond the restrictive figures of women that simplistically divide actual women into two kinds— good girls, loving mothers, and adoring, nonthreatening sisters on the one hand, and manipulative mistresses, man-hating lesbians, and psychotic dropouts on the other—is inseparable from resistance to the kind of moralizing feminism that tries to distinguish good feminism from so-called bad feminism. The classic example of this kind of moralizing is the claim made by antipornography feminists that women who disagree with them are "bad" rather than just wrong. The language of political disagreement yields to charges of evil doing. This form of "guilt tripping" impedes the political battles of feminism.[3]

Another dimension of this moralizing that makes it specifically dangerous for how women see themselves is that it reinstates and unconsciously reinvests with validity what Jacques Lacan calls the "psychical fantasy of woman" (see *Feminine Sexuality*). For Lacan, in the place of the mature female subject, which would demand that the feminine be given a symbolic reference, are objectified fantasies of our sex. Given the masculinization of what Lacan calls the symbolic order, in which the referent for a male masculine identity is the phallus and the symbolic order itself is engendered in and through the law and linguistic encoding unconsciously associated with the imaginary father, the mother/Other is barred from conscious registration. For Lacan, this bar, which prevents us from symbolizing the ultimate object of desire, splits women into two main categories of objects. Following Melanie Klein, the split is between "good" mothers and "bad" mothers (306–43). These figures of the good and bad woman implicate this fundamental splitting and lead to their simplistic description. I have argued elsewhere that the psychical fantasy of woman is inseparable from the semantics of closure encoded in and through gender hierarchy. Thus, my addition to Luhmann is that, as feminists in particular, we need the ethical in order to "switch off" the moralizing that is inseparable from the enforcement of ladylike behavior. The virtues associated with what it means to be a "lady" are themselves part of a moral system and we should expose them as such. We certainly need to limit this moral system precisely as it limits our freedom. The "manners" imposed on the "ladies" are ultimately our chains.

The moralizing of feminism that tries to divide us into good feminists and bad feminists unconsciously replicates and reenforces the system of morality imposed upon women. The ethical, then, is against the restrictive femininity we endure because of the very idea of the "good girl." Thus, for my own feminist reasons, I, too, warn against the market of approval inevitably associated with morality.

For both Luhmann and myself—even though I do not completely endorse Luhmann's own historical understanding of the ethical—the ethical is itself a paradox. For me, the specific paradox of the ethical marks it as a limit principle. We need, then, to return to why I justify the ethical and justice as limit principles. But to do so, I need first to offer my own reading of Wittgenstein based primarily on the later Wittgenstein and not the Wittgenstein of the *Tractatus*. This reliance, however, does not mean that there are two neatly discernible Wittgensteins. Instead, there are questions, and re-posings of those questions, that endlessly haunted the man, including, as Rasch points out, the problem of the impossibility of ethics itself.

On my reading of the later Wittgenstein, the frame that marks the world as "what is" cannot be *conceptualized as* an absolute limit. We can only know this divide between "what is inside" and "what is outside" through metaphor. The boundary of the form of a life is a metaphor, in other words, and not a concept. Like all metaphors, the excess inherent in the identification through transference points beyond itself. Thus, paradoxically, the limit recedes before its own linguistic expression. We know the limit only against what it demarcates, not in itself. Luhmann makes the point that this is true of all systems, not just linguistic systems. Whenever we try to describe the ethical as a full system of ethics, for example, we will always be returned to the paradox that it can only be understood in and against the very moral systems that it limits. Therefore, the ethical marks the limit of morality demanded by morality itself. Justice, in like manner, marks the limit of law, particularly in a modern legal system, demanded by the legal system itself. But I will return to the specific sense in which I use "justice" and the "ethical" shortly. For now, I want to tackle Rasch's own reading of Wittgenstein, which beautifully describes Wittgenstein's early essay on ethics, in which the ethical is marked by the consistent and persistent elaboration of its own impossibility.

This impossibility of the ethical followed from Wittgenstein's philosophical project in the *Tractatus,* which sought to establish linguistic form as a self-identical form which could, in fact, give us a full description of the world as it is since it would provide us with the self-enclosed semantic system in which all valid propositions could be encompassed. The world would be these propositions—nothing more, nothing less. Therefore, by definition, there would be nothing outside this semantic system understood as a self-identical form of which we could make sense. But what Wittgenstein came to realize was that the boundary of language, which had to be conceptualized in order to philosophically explain the enclosure of the semantic system as a self-identical form, could not, in fact, be reached by philosophical knowledge. His philosophical conclusion was that a form of life or a language game could never be *known* as a self-identical form.

But there was an even greater paradox for Wittgenstein, at least in my reading of him: it is through the very attempt to conceptualize the boundary that encloses a semantic system that we run up first against the limit of philosophical justification and then of sense itself. We run up against the limit precisely as it recedes before our attempts to adequately philosophically conceptualize it. But how can one run up against that which recedes when we try to describe it? How can the limit be there when it cannot be known?

In my reading, Wittgenstein knew that he could not use a philosophical explanation to illustrate this paradox, which is why he rejected traditional philosophical writing in his later work *(Philosophical Investigations).* Yet it was precisely this irresolvable paradox that was the limit for Wittgenstein of philosophical justification. His highly original style, which demonstrated the inevitability of running into this paradox, was adopted because it was the only style adequate to the task of evoking the limit of language that both operates to demarcate our life world, yet recedes before our attempts to conceptualize it.

For me, however, there is an unerasable moment of utopianism in the endless demonstration of this paradox in both Wittgenstein and Jacques

Derrida. It is the significance of this particular form of utopianism that is crucial for feminism. The challenge to any attempt to philosophically secure the bounds of meaning implicitly defends the possibility that we operate within an ever wider field of meaning without knowing where that field ends. As the boundaries recede, we have more space to dream and reimagine our form of life. The very impossibility of knowing the boundaries that guarantee meaning is unsettling if one seeks security in an established world of sense. As feminists know only too well, however, we have been tied down by the bonds of the meaning of femininity. The very impossibility of knowing the limit opens up an endless horizon precisely because we cannot tell philosophically what is inside and what is outside. In the end, then, "the beyond"—or what I mean by "the beyond"—is just this paradox. We cannot have a full description of the world as it is. Thus, the philosophy of the limit is not seduced into transcendence; it instead marks what is "beyond within" as utopian possibility.

In Derrida, this "beyond within" is given a much more explicit ethical recognition than it is in Wittgenstein. I am, of course, aware that my reading of Wittgenstein is not the received one, though I believe it is one consistent with his own relentless philosophical demonstration of the paradox inherent in the demarcation of a life world and language game.

For Luhmann, systems are also demarcated by their other. But this other is just a form of self-limitation of the system. The other is *of* the system, not the other *to* the system. The rejection of the philosophical demonstration of the limit of the system as the mark of the other *to* the system is explicitly recognized by Luhmann as his difference from Derrida. To quote Luhmann: "Information is, according to Gregory Bateson's oft-cited dictum 'a difference that makes a difference'" (54). Regardless of what one thinks of their ontological and metaphysical status or their inclination ascript (Derrida) as similar approaches, differences direct the sensibilities that make one receptive to information. Information processing can only take place if beyond it, something has been experienced as this way and only this way, which means that it's been localized in a framework of differences. The difference functions as a unity to the extent that it generates information, but it does not determine which pieces of information are called for and which patterns of selection they trigger off. Differences, in other words, do not delimit a system; they specify and extend its capacity for a self-delimitation. For Luhmann, this difference is one that can be philosophically decided.

I have argued that for both Derrida and Wittgenstein, on the other hand, it is ultimately philosophically *undecidable* whether a system is

104   Drucilla Cornell

delimited by its other or whether it is a system that delimits itself by its other only as a process of self-limitation. I have made the further argument that since Luhmann himself always turns us to the observer and since the difference between the self-limitation of the system and the system as delimited by its other as a matter of systems operations cannot be philosophically decided, we will be turned to how the observer sees the system. Of course my additional argument here is that the feminist—as placed outside the symbolic order—sees the system, including the system of gender, as a system and that its delimitation is exactly what puts us outside of it.

But on a more philosophical plane, it is of course correct, at least in my reading of Derrida, that he provides a quasi-transcendental analysis that shows why, if the system is self-limited, it is necessarily delimited by its other. Thus, as well as having the "beyond within" the real and the challenge to the very conceptualization of the immanent and the transcendent as an absolute divide, Derrida also challenges the difference between delimitation and self-limitation because the only system that could be truly self-limiting and only self-limiting would, by definition, have to encompass all other systems. Thus, since systems operate, even in Luhmann, by delimiting themselves from the other, there will always be the other who, as observer, could see the system both as a system and as delimited by virtue of their very outside or marginalized position against that system.

But why have I argued in *The Philosophy of the Limit* that this quasi-transcendental moment is ethical? For me, it is more precise to argue that it is the "beyond within the real" and the delimitation of the system by its other that keeps open the space for the ethical and political challenge to what "is" because what "is" is never simply there. This space is not literal space but the metaphoric indication of the beyond. In other words, this quasi-transcendental moment need not itself be ethical, but it is the space of both the ethical relation and a new understanding of the political. In the terms that Rasch uses to describe the project of the early Wittgenstein in his essay on ethics, the paradox here would be that it would be the impossibility of showing the impossibility of the ethical that maintains the space for the ethical itself. Derrida, in my reading, heeds the ethical in precisely this sense. What I am calling the space of the ethical, Derrida has explicitly addressed as the condition of repoliticization. To quote Derrida:

> Permit me to recall very briefly that a certain deconstructive procedure, at least the one in which I thought I had to engage, consisted from the outset in putting into question the onto-theo- but also archeo-teleological concept

of history—in Hegel, Marx, or even in the epochal thinking of Heidegger. Not in order to oppose it with an end of history or anhistoricity, but, on the contrary, in order to show that this onto-theo-archeo-teleology locks up, neutralizes, and finally cancels historicity. It was then a matter of thinking another historicity—not a new history or still less a "new historicism," but another opening of event-ness as historicity that permitted one not to renounce, but on the contrary to open up access to an affirmative thinking of the messianic and emancipatory promise as promise: as *promise* and not as onto-theological or teleo-eschatological program or design. Not only must one not renounce the emancipatory desire, it is necessary to insist on it more than ever, it seems, and insist on it, moreover, as the very indestructibility of the "it is necessary." This is the condition of a repoliticization, perhaps of another concept of the political. (74–75)

This "it is necessary" is what I refer to, in *The Philosophy of the Limit*, as the "beyond" within the real itself as both the space for the ethical, and I agree with Derrida, ultimately, for the space for a different understanding of the political. The emancipatory desire, then, does not have to be justified or warned against as a temptation to transcendence.

But I do want to note that I have become critical of my own appropriation of Immanuel Levinas, on which I rely in *The Philosophy of the Limit*, to represent the ethical as the beyond within the real (171–72). It is beyond the scope of my remarks here to fully elaborate why I am critical of my earlier formulation, but I do want to note that it has to do with what Rasch calls the anthropomorphization of the outside as the other. In Levinas, we are always called by the other to an infinite responsibility. It is this other as the confrontation with the human face that calls us. This, of course, is the other as both other to the system, as Rasch rightly remarks, and as within it. This, he believes, is what allows me to say that the ethical is in reality. My own critique of my reliance on Levinas is ultimately that his own metaphorization of the ethical relationship is contaminated by the very sentimentality of the images of feminine sacrifice he uses to represent it. I, of course, remark on and critique that contamination in *The Philosophy of the Limit* (91–115), but on rethinking my own position, I have concluded that this contamination is inseparable from deradicalization of what I have called "the space for the ethical" by trying to imagine it in the way that Levinas does. Thus, I now would insist on the space of the ethical as the space of the "it is necessary" in what I call the quasi-transcendental analysis of both the delimitation of the system and with it, the deconstruction of the divide between the immanent and the transcendent.

Within law, more specifically, my argument is that justice should be understood as a limit principle in precisely the paradoxical sense that any attempt to turn justice into a fully realized conception will always have justice as the warning against the hubris that it has finally been realized. This is justice now serving a similar, but not identical function, to the one that Luhmann allots to ethics in moral systems. For me, specifically, justice as the demand for fair evaluation always carries within it the injustice that any form of evaluation can never fully comprehend the singularity of each one of us as a person. Yet, within a democratic, modern legal system, I also argue that we are called upon to fairly evaluate each one of us as a person.[4] Thus, there is, in the very call for fair evaluation, an irresolvable tension. I use "person" in *The Imaginary Domain* in the classic Latin sense to be that which "shines through" and marks our singularity. To become a person is a project. I reject the assumption of the person as a given. But, in spite of my insistence that the ultimate injustice is to assume that one has achieved justice, at least justice defined as the full recognition of our singularity as persons, I also call for the demand imposed by a modern, legal system that we, as law professors, lawyers, and judges endlessly try to synchronize the basic values of our legal system as justly as possible. Thus, the call for justice always evokes what Derrida heeds as the "it is necessary," even though it remains as a promise and a command.

For me, a truly nonviolative relationship to the other would involve this recognition of the other as absolutely unique and singular and this no legal system can ever achieve, precisely because it is called to make fair evaluations. It is this nonviolative relationship to the other that I call the ethical relationship. In spite of the divergence from Kant in the sense that I, like Luhmann, deconstruct the absolute divide between the noemenal and the phenomenal and therefore have sought to redefine the definition of the person away from that divide, the call to the nonviolative relationship to the other clearly has Kantian overtones (see *Imaginary Domain*).

In terms of the law itself, the Kantian understanding of the nonviolative relation to the other translates into the legal recognition that none of us should be reduced to an object instead of being recognized as a person. Before the law, this has led me to justify what I call the "degradation prohibition," which demands that, within our legal system, none of us should be regarded on a hierarchical scale of being that degrades some of us as unworthy of personhood. This degradation prohibition can be understood to inhere in a legal system that has moved from hierarchical, stratified differentiation to functional differentiation, in which none of us is marked as inherently scaled down because of our station in life. This shift, Luhmann argues, is the hallmark of a modern legal system. Of course, Luhmann would be very skeptical of this Kantian language, but my argument here is that this language inheres in our legal system itself, with its unflagging democratic pretensions and with the demand that our legal decisions not only be right, but just. Thus, such an argument does not appeal to what Luhmann would call moral categories. It appeals instead to the idea of the equivalent, legal personhood that inheres in the modern legal system and that has been best elaborated for purposes of a conception of law by Kant in his later political writings.[5]

My tribute in *The Philosophy of the Limit* to Justice Blackmun did not focus on his attempt to recognize the other as other in all her singularity. Instead, I argued that Blackmun heeded the call of justice in his attempt to recognize women as equal persons, as this ultimately implicated the legal protection of abortion as a right. But my point was not, as Rasch writes, to argue that Justice Blackmun sought to recognize the ethical relationship as much as it was that he sought to heed the call to justice, a call that took him into territory that had been unchartered in our own constitutional schema. My argument was that *Roe v. Wade* had "doctrinal difficulties" precisely because it demanded that he heed the call of justice for women and attempt to chart out what had remained unchartered territory.

Thus, I disagree with Rasch that I was collapsing the moral into the legal. I was arguing, instead, that Justice Blackmun, in heeding the call for justice, was exemplary in exercising his judicial responsibility and doing so by proceeding through the legal concept of privacy itself. Even if I argue, and I do, that privacy is an inadequate legal justification for abortion, the inadequacy was not a failure of Blackmun's judicial intelligence but a failure of the legal concept of privacy itself to encompass what was at stake—the bodies of women—in the debate over abortion. Of course, I recognize that since a legal system always wields extraordinary violence, the vindication of the right of abortion can itself be read as the recognition of a form of necessary violence, particularly if one believes the fetus to be a form of life. Still, within our legal system, we have not defined the fetus as a person. I accept that definition.

To conclude, ethical feminism, as I define it, does not succumb to the temptations of transcendence. Instead, feminism thrives in what I call the future to come, which is already in the process of arriving and which is necessarily there as a promise of a different world.

Notes

1. Drucilla Cornell, *Philosophy of the Limit* (New York: Routledge, 1992).

2. Please see my discussion of Luhmann in "The Relevance of Time to the Relationship between the Philosophy of the Limit and Systems Theory: The Call to Judicial Responsibility," in *Philosophy of the Limit.*

3. See MacKinnon, where she charges that women who disagree with her are collaborators with the pornographers.

4. Drucilla Cornell, *Imaginary Domain* (New York: Routledge, 1992).

5. See Immanuel Kant, *Kant: Political Writings,* ed. Hans Reiss (Cambridge: Cambridge University Press, 1970).

Works Cited

Cornell, Drucilla. *The Imaginary Domain: A New Perspective on Abortion, Pornography, and Sexual Harassment.* New York: Routledge, 1995.

———. *The Philosophy of the Limit.* New York: Routledge, 1992.

Derrida, Jacques. *Specters of Marx: The State of Debt, the Work of Mourning, and the New International.* Translated by Peggy Kamuf. New York: Routledge, 1994.

Kant, Immanuel. *Kant: Political Writings.* Edited by Hans Reiss. Cambridge: Cambridge University Press, 1970.

Klein, Melanie. *Love, Guilt and Reparation and Other Works, 1921–1945.* New York: Free Press, 1975.

Lacan, Jacques. *Feminine Sexuality: Jacques Lacan and the École Freudienne.* Edited by Juliet Mitchell and Jacqueline Rose. Translated by Jacqueline Rose. New York: Pantheon, 1982.

Luhmann, Niklas. *Love as Passion.* Translated by Jeremy Gaicresard and Doris L. Jones. Cambridge: Harvard University Press, 1986.

MacKinnon, Catharine. *Only Words.* Cambridge: Harvard University Press, 1993.

Rasch, William. "Immanent Systems, Transcendental Temptations, and the Limits of Ethics." In *Observing Complexity,* edited by William Rasch and Cary Wolfe. Minneapolis: University of Minnesota Press, 2000.

Wittgenstein, Ludwig. *The Philosophical Investigations.* Translated by G. E. M. Anscombe. Oxford: Basil Blackwell, 1968.

———. *Tractatus Logico-Philosophicus.* Translated by C. K. Ogden. New York: Routledge, 1981.

II

OF REALISM AND RECURSIVITY:
SYSTEMS THEORY AND THE
POSTMODERN EPISTEME

6. Theory of a Different Order: A Conversation with Katherine Hayles and Niklas Luhmann

Niklas Luhmann, N. Katherine Hayles,
William Rasch, Eva Knodt, and Cary Wolfe

This discussion was conducted September 21, 1994, at the Institute for Advanced Study at Indiana University, Bloomington, where Niklas Luhmann was a guest Fellow for two weeks. Both Luhmann and N. Katherine Hayles were participating in a conference at the university later that week, organized by William Rasch and Eva Knodt, entitled "Systems Theory and the Postmodern Condition." As a basis for discussion and exchange, before the interview Hayles was given a copy of Luhmann's essay "The Cognitive Program of Constructivism and a Reality That Remains Unknown," and Luhmann was provided with a copy of Hayles's "Constrained Constructivism: Locating Scientific Inquiry in the Theater of Representation." The conversation was organized and moderated by William Rasch, Eva Knodt, and Cary Wolfe.

CARY WOLFE: I'd like to begin with a general question. In your different ways you have both explored a second-order cybernetics approach to the current impasse faced by many varieties of critique. And that impasse, to schematically represent it, seems to be the problem of theorizing the contingency and constructedness of knowledge without falling into the morass of relativism (as the charge is usually made) or, to give it a somewhat more challenging valence, without falling into philosophical idealism. You both have worked on this, and I'm wondering if each of you could explain, in whatever order you'd like, what makes second-order theory distinctive, and how it might help move the current critical debates beyond the sort of realism versus idealism deadlock that I've just described.

KATHERINE HAYLES: Would you care to go first?

NIKLAS LUHMANN: OK. Well, I reduce the general term "second-order" to second-order observing, or describing, what others observe or describe. One of the distinguishing marks of this approach is that we need a theory of observation which is not tied to, say, the concept of intelligence, the mind of human beings, but a more general theory of observation that we can use to describe relations of social systems to each other, or minds to social systems, or minds to minds or maybe bodies, to neurophysiological systems, or whatever. So, it needs to be a general theory of observing—and I take some of these things out of *The Laws of Form* of George Spencer Brown—to think of observing as an operation that makes a distinction and is then bound to use one side of the distinction, and not the other side, to continue its observations. So we have a very formal concept of observation. And the problem is then, if you link different observing systems, what can be a cause of stability, how can—in the language of Heinz von Foerster and others—*eigenvalues,* or stable points or identities, emerge that both sides of a communication can remember? And I think this is the idea which goes beyond the assumption that relativism is simply arbitrary: every observation has to be made by an observing system, by one and not the other, but if systems are in communication, then something emerges which is not arbitrary anymore but depends on its own history, on its own memory.

KH: For me, second-order theory would be distinct from first-order theory because it necessarily involves a component of reflexivity. If you look at first-order cybernetics, it's clear that it has no really powerful way to deal with the idea of reflexivity. In the Macy conference transcripts, reflexivity surfaced most distinctly in terms of psychoanalysis, which was threatening to the physical scientists who participated in the Macy conferences because it seemed to reduce scientific debate to a morass of language. When they would object to Lawrence Kubie's ideas, who was the psychoanalyst there, he would answer with things like "Oh, you're showing a lot of hostility, aren't you?" To them, that was almost a debasement of scientific debate because it kept involving them as people in what the conference was trying to do. There were strong voices speaking at that conference in favor of reflexivity—people like Gregory Bateson and Margaret Mead—from an anthropological perspective. But because reflexivity was tied up with psychoanalysis and the complexities of human emotion, it seemed to most people at the Macy conferences simply to lead to a dead end. When Maturana and Varela reconceptualize reflexivity in *Autopoiesis and Cognition,* they sanitize reflexivity by isolating the observer in what they call a "domain of description" that remains separate from the autopoietic processes that constitute the system as a system. I think Professor Luhmann's work is an important refinement of Maturana's approach because he has a way to make the observer appear in a non–ad hoc way; the observer enters at an originary moment, in the fundamental act of making a distinction. Nevertheless, I think that the history I've just been relating is consequential—the point that you can get to is always partly determined by where you've been. The history of second-order cybernetics is a series of successive innovations in which the taint that reflexivity acquired through its connection with psychoanalysis has never completely left the theorizing of the observer as it appears in that tradition. This is quite distinct from how reflexivity appears in, say, the "strong program" of the Edinburgh School of Social Studies of Science, where they acknowledge that the act of observation is grounded in a particular person's positionality.

Reflexivity has been, of course, an ongoing problem in both science and the history of science. When reflexivity enters relativity theory, for example, it has nothing to do with a particular person's personality, cultural history, or language; it has only to do with the observer's physical location in space and time. Relativity theory is not reflexive, it is only relative. To try to arrive at a theory of reflexivity which would take into account the full force of the position of the observer, including personal history, language, a culture, and so forth, has been, I think, a very important and extremely difficult problem to solve. To me, it's essential to talk about the observer in terms that would take account of these positional and locative factors as well as the theoretical question of how is it that we can know the world.

cw: To what extent do you think that in their recent work, Maturana and Varela have tried to move in this direction? I'm thinking now of the collaboration of Varela and Thompson and Rosch in *The Embodied Mind,* but more broadly of the whole concept of embodiment in second-order cybernetics, which has certain affinities with Donna Haraway's work on this problem, which is very much in the register that you were emphasizing. I'm thinking, too, of the explicit derivation of an ethics at the end of *The Tree of Knowledge* from second-order cybernetics. To what extent, then, do you see much of this work moving in that direction, and if so, is it moving in the way that you would like?

kh: You know, it's difficult to try to coordinate all these works, because they seem to me all significantly different, maybe because I'm geared to thinking about texts, and therefore about the specific embodiment of these ideas in the language they use. But to compare just for a moment *Autopoiesis and Cognition* with *The Tree of Knowledge*: in the latter the authors write for a popular audience, and in the process the work changes form. It goes from an analytical form into a circular narrative. And with that shift come all kinds of changes in their rhetorical construction of who the observer of that work is, as well as of themselves as observers of the phenomena that they report. In this sense, *The Tree of Knowledge* is more positioned. But it does not solve a problem also present in *Autopoiesis and Cognition*—that is, using scientific knowledge to validate a theory which then calls scientific knowledge into question. I'm thinking here specifically of "Studies in Perception: Reviews to Ground a Theory of Autopoiesis." Autopoiesis leads to a theory of the observer in which there is no route back from the act of observing to the data that was used to generate the theory in the first place. The problem is exacerbated in *The Tree of Knowledge*. Even as they move from a "domain of description" to a more capacious idea of a linguistic realm in which two observers are able to relate to each other, there arise other problems having to do with the work's narrative form.

WILLIAM RASCH: What is your reaction to this?

NL: Well, there are several reactions. One is that I have difficulties, regarding the later work, comparing Maturana and Varela. Maturana advanced in the direction of a distinction between the immediate observer and the observer who observes another observer. The "objective reality" is that there are things, or niches, which are not reflected in the immediate observer's boundaries. But on the other hand, if you observe that observer, then you see how he or she sees the world by making this distinction. But the limit of this type of thought is the term "autopoiesis" itself as a system term. Autopoiesis was another term for circularity, that was its beginning. Maturana talked about cells in terms of circular reproduction and then, after some contact with philosophers, used "autopoiesis," finding the Greek term more distinctive. But there remains in Maturana the idea that circularity is an objective fact, and so the problem of self-reference is not really confronted in the theory—not in the sense of, for example, the cyberneticians who would say that a system uses its output as input and then becomes a mathematical cosmos with immense amounts of possibilities which cannot be calculated anymore, as in Heinz von Foerster or Spencer Brown's discussion of a "re-entry" of the distinction into the distinguished. And there are, within these more mathematical theories, possibilities which are not visible, I think, in the writings of Maturana and Varela. They are too empirically tied to biology. And then of course we have always this discussion of whether one can use biological analogies in sociology or in psychology or not, which doesn't lead anywhere.

WR: I have a question. Professor Luhmann, you said that you wanted to find a definition of observation that is on a very formal basis, that does not only apply to consciousness, but to systems of all sorts. When you, Professor Hayles, talked about observation, the sense of an individual came out more because you were talking about the person's locality, the observer's situation. Do you have a sense that observation is tied strictly to consciousness? Or is observation also for you a more formal definition that can be applied to systems other than consciousness?

KH: For me, observation is definitely tied to consciousness. In Professor Luhmann's article "The Cognitive Program of Constructivism and a Reality That Remains Unknown," you have a paragraph where you're talking about the observer, and you list a series of things like a cell, a person, and so forth. On my own copy of that article I put a big question mark in the margin: can a cell observe? Of course, I realize that it's partly a matter of definition, and you're free to define the act of observation however you want. But, for me, a cell could not observe in the way I use the term.

EVA KNODT: Could you maybe clarify . . .

WR: Let's let Professor Luhmann clarify how a cell can observe.

NL: Well, it makes distinctions. It makes a distinction with input/output, what it takes in or what it refuses to take in, or a distinction about its own internal reproduction, to do it in a certain way and not in another way. I'm not sure whether making distinctions implies the simultaneity of seeing both sides, or whether it is just discrimination. The immune system discriminates, of course, but does it know against what it is discriminating? And if you require for a concept of observing that you see both sides simultaneously, and the option becomes an option *against* something, then I would not say that cells are observing or immune systems are observing. They just discriminate. But for me this is not very important. It would be very important for Maxwell's Demon, for example, that he can distinguish—or it, whatever it is, can distinguish—what belongs on which side. But it is hardly thinkable for us, because we are always using meaning in constructing reality. So the problem is to think of distinction, of observation, without the idea of seeing out of the eyes, out of the corner of the eyes, the other thing which we reject or give a negative value. So we, psychologically and socially, use the idea of meaning, so that "observing" becomes a distinct characteristic. And there is a question, of course, of whether we should extend it. But this is I think a terminological . . .

EK: I have a follow-up here, because I also was puzzled in the beginning when I started reading your work about this use of observation, and how it is different from this metaphorical idea that one thinks one sees with the eyes. It's very hard to separate oneself from it. Where exactly do you see the advantage of widening this concept of observation to an extent that it is no longer located in consciousness?

NL: For me, the advantage is to make possible a kind of interdisciplinary commerce, a kind of transference of what we know in cybernetics or biology into sociology or into psychology. Saying that there are very general patterns which can just be described as making a distinction and crossing the boundary of the distinction enables us to ask questions about society as a self-observing system. What happens in a self-observing, self-describing system? This is not only a question for conscious systems. I mean, there are five or more billion conscious systems, and you cannot make any theory of society out of adding one to another or dissolving them all into a general notion like the transcendental subject. But you can make some headway, perhaps, by using the formal idea of observing, and of making distinctions, to understand a system that has a recursive practice of making distinctions and guiding its next distinctions by previous distinctions, using memory functions, and all this. There are formal similarities between psychic systems and social systems, and this is for me important in trying to write a theory, a social theory, of self-describing systems, in particular of society.

WR: Shall we move on to a topic that is perhaps broached more directly in the two articles, and that is the topic of reality. Based on your reading of each other, how would each of you distinguish your notions of reality from the other? Both of you use the term *reality,* and yet strict realists would not recognize the term as each of you use it. But how do you observe each other using the term *reality*? Either one of you start.

KH: I'll be glad to start. In Professor Luhmann's article I alluded to before, the sentence that I found riveting was this: "Reality is what one does not perceive when one perceives it." It was when I got to that sentence that I thought I was beginning to grasp his argument because I fully agree with that, with one important reservation. I, too, agree that whatever it is that we perceive is different, dramatically different, than whatever is out there before it is perceived. If you want to call what is out there before it is perceived "reality," then we do not perceive it, because the act of perception transforms it. Where I would differ is with the distinction between reality and nonreality, the binary distinction which he uses so powerfully in a theoretical way. I am concerned about a fundamental error that has permeated scientific philosophy for over three hundred years: the idea that we know the world because we are separated from it. I'm interested in exploring the opposite possibility, that we know the world because we are connected to it. That's where I would distinguish the approach I take. It is not really even a disagreement; it's more a matter of where you choose to put the emphasis. Do you choose to emphasize the interfaces that connect us to the world, or do you choose to emphasize the disjunctions that happen as distinctions are drawn?

cw: Professor Luhmann, I imagine you would like to respond . . .

nl: This formulation has a kind of ancestry, and in former times was associated with the idea of existence, with the idea, to put it another way, that I see trees, but I don't see the reality of trees. And if reality refers to *res,* and *res* is the thing, then you have visible and invisible things—and that's the world. In this philosophical tradition, the problem simply was not possible to formulate. But the formulation that reality is what you don't see if you see something can be phrased in different ways. And one of these other possibilities is to say that reality emerges if you have inconsistency in your operations; language opposes language, somebody says yes, another says no, or I think something which is uncomfortable given my memory, and then you have to find the pattern of resolution. Reality is then just the acceptance of solutions for inconsistency problems, somewhat as, in a neurophysiological sense, space is just produced by different lines of looking at it, by internal confusion and then a solution to the internal confusion, which is in turn produced by memory that could not remember if it could not make differences in time. I am here now, but before I was in the hotel, and before that I was in the restaurant, and were this everything at the same moment, then I could not have any kind of memory. So time is real because it tries to create consistency and solve inconsistency problems. And this explains why reality is not an additional attribute to what you see, but is just a sign of successful solutions. This also helps us to see the historical semantics of reality. For example, "culture" at the end of the eighteenth century is a term which is able to organize comparisons—regional ones (French, German, and so on, or Chinese or European) and historical ones—so that there is a new pattern, some striking insight that is possible because the compared things are different. And "reality," as a result of functional comparisons, is just this kind of insight. You needn't have a more abstract notion of culture or identity or society, or whatever, to be able to handle contradictions which otherwise would obstruct your cognition.

cw: Let me just ask, for clarification, is this reality to which you are referring here different from the reality which is a kind of a creation or accumulation of what you elsewhere call *Eigenvalues,* or is that in fact what you are describing?

nl: No, I think that is just another formulation.

cw: OK, all right. I'd like to come back to something you said, Professor Hayles, and ask you about this issue of connection versus separation that you're interested in. One of the things that's distinctive to me about second-order cybernetics—its central innovation, I think—is that it theorizes systems that are both closed *and* open: in Maturana and Varela, the attempt to theorize closure on the level of operations or organization, but openness to the environment on the level of structure. So, in a sense, isn't that a theory of self-referential systems which are nevertheless connected to the reality in which they find themselves?

kh: Well, for Maturana and Varela systems are connected by structural coupling. What that gets you in explanatory power is a way to explain the plasticity of systems and changes in structure. Where I have a fundamental difference with Maturana and Varela is in their assumption that there is no meaningful correlation between stimuli that interacts with receptors and information that the receptors generate. This may finally come down to religious dogma; I am of one faith and they are of another. I have studied the articles on perception which Maturana and his coauthors published on color vision in humans and on the visual system of the frog. I do not believe the data support his hypothesis that there is no correlation between inside and outside. It was a bold and courageous move to make that assumption, because it allowed them to break with representation and to avoid all of the problems that representation carries with it. It did get them a lot of leverage. But it's one thing to say there is no correlation, and another to say that the transformations that take place between the perceptual response and outside stimulus are transformational and nonlinear. The latter, I believe, is more correct than the former. I think it's important to preserve a sense of correlation and interactivity. This is primarily where I differ from them.

WR: Could I just follow up, and then maybe both of you could comment. You mentioned before that where you had differences, when you were speaking of Professor Luhmann's work, was with the assumption that knowledge of the world is attainable because of separation from the world. If now you're saying that there is some way of thinking of a correlation between an outside and an inside, doesn't that ontologize separation from the world, and doesn't that get you back into what you were trying to get out of—that is, the idea that we can only know the world because it is outside of us and it has causal effects on us through sensory perception? Doesn't that solidify the inside/outside distinction? Why not talk instead about closure and knowledge coming from the inside, where the inside/outside distinction is made in the inside, and there is thus a more fluid relationship between the two, where you know the world because you are the world?

KH: Well, if you allow the distinction to fall into an inside/outside, as it certainly can, then you're back essentially to realism in some form and also representation. What I was trying to do in my article on constrained constructivism was to move the focus from inside/outside into the area of interaction, where inside and outside meet. That precedes conscious awareness, but it is in my view an area of interaction in which, precisely, a correlation is going on between stimuli and response. So . . .

EK: Could you elaborate a bit? I have a problem here because you said a little earlier that whether or not you accept the idea of closure comes down to dogma or faith, and now you're referring to some observations that seem to confirm the model that you're proposing. Could you say a little bit more about what kind of evidence leads you to your particular choice?

KH: If we start from the frog article, which was the beginning for Maturana, what the article concludes (this is a near quotation) is that the frog's eyes speak to the frog's brain in a language already highly processed. It does not, however, show that there is no correlation between the stimuli and the response.

EK: Yes, but what is the status of this correlation? I mean, that's what the observer constructs as the frog's reality.

KH: Yes, that's right. That is, what is constructed is the frog's reality.

EK: From the human point of view.

KH: Yes. From the experimental point of view, to be more precise.

NL: But then you have the question, who is the observer? If it is a scientist, he or she can make theories and can see correlations, but if it's a frog itself, then things are different. Maturana talks about structural couplings and so on, but the frog as such constructs his reality as if it were outside, to solve internal conflicts. So, in this sense, the question is, why does a closed system like a brain need a distinction inside/outside to cope with its own problems, and why does it construct something outside that externalizes the internal problems of the workings of the brain, just to order his world, in which he himself is, of course, given?

WR: Can I follow up on that? This brings us to the notion of consistency, which Professor Hayles talks about in her article. And if I understand that correctly, the fact is that each one of us in this room would probably open that door to try to walk out of this room. We're all constructing the world based on internal contradictions, but it all happens to be the same world with reference to this room and these five people. How is that possible?

NL: Well, I think it would be—to take an example from the article of Professor Hayles—that if we jump out of the window we would contradict our own memory. We have never seen someone stop before they hit the ground, so we simply sort out our contradictions, as long as it is not necessary to change it, within the old pattern. So we go through the door and take the elevator, and this is reality as a solution of formal contradictions. Maybe we try once to jump from too high a place, but we never see apples or something stopping in the middle of the fall.

WR: So it's strictly experiential?

NL: It is just the solution of an internal conflict of new ideas or of variations within your memory.

WR: So in a sense, you both believe in constraints. If I understand you correctly now, Professor Luhmann, you would phrase constraints in terms of internal operations, especially memory, in this case. How do you, Professor Hayles, see the constraints that would prevent us from walking out of this window or trying to walk through that door? If you don't want to be a realist, and say because it's a door or because of gravity, how do you define what the constraints are?

KH: Well, the way I think about it is that "reality" already carries the connotation of something constructed, so I prefer to use the term "unmediated flux." The unmediated flux is inherently unknowable, since by definition it exists in a state prior to perception. Nevertheless, it has the quality of allowing some perceptions and not others. There is a spectrum of possibilities that can be realized in a wide range of different ways, depending on the perceptual system that's encountering them, but not every perception is possible. Therefore there are constraints on what can happen. We can all walk out the door together because we share more or less the same perceptual system—more importantly because we share language, which has helped to form our perceptual systems in very specific ways.

WR: How does that differ from memory as Professor Luhmann described it? In other words, I'm being very devious here in trying to coax the word "physical" out of you. How can you describe what you're describing without using the term "physical constraints"? Or are the constraints strictly in the way the brain is structured?

KH: I believe there are constraints imposed by our physical structure; I have no doubt of that. I think there are also constraints imposed by the nature of the unmediated flux itself.

WR: What one would conventionally call the actual physical structure of the unmediated flux?

KH: Yes, that's right.

NL: Then, if you use for a moment the idea that reality is tested by resistance—that's Kant—how can you have external resistance if you cannot cross the boundary of the system with your own operations? You cannot touch the environment with your brain, and even if you touch it you feel something here [points to his head] and not there, and you make an external reality just to explain that you feel something here [points again] and not in other places on your body. So, finally, it's always an internal calculation, otherwise you should simply refuse the term "operational closure." But if we have operational closure we have to construct every resistance to the operations of a system against the operations of the same system. And reality then is just a form—or, to say it in other terms, things or objects outside are simply a form in which you take into account the resolution of internal conflicts.

EK: If that model holds, can you account for the historical emergence of this idea that there is, and ought to be, a difference between the reality as unmediated flux—what we do not perceive when we perceive—and the world of objects that we encounter in everyday life? I mean, does this idea itself have a similar function?

NL: I'm not sure . . .

EK: Starting with Kant, we find the distinction between the unknowable noumena and phenomena, where you locate some sort of reality outside and then you talk about constructed phenomenological reality. Could one apply this idea that you just mentioned—that reality has the function of neutralizing contradictions—to account for the emergence of this historical distinction?

NL: The emergence of this kind of internal distinction between inside and outside is even earlier. A system makes a distinction because it couples its own operations to its environment over time and has to select fitting operations, or it simply decays. Then, if it makes such a distinction, it has no way to handle the environment except by reconstructing or copying the difference between system and environment into the system itself, and then it has to use an oscillator function to explain to itself something either as an outcome of internal operations or as the "outside world." In Husserl it's clearer than in Kant, that you have noesis and noema, and you have intentions, and you can change between the two and put the blame on your own thinking or be disappointed with the environment. And to explain how our system copes with this kind of distinction, instead of just checking out how it is out there, we need an evolutionary explanation of how systems survive to the extent that they can learn to handle the inside/outside difference within the system, within the context of their own operation. They can never operate outside of the system.

WR: Do you have a response . . . ?

KH: This is not really so much a response to the thought that Professor Luhmann was just developing as a more or less independent comment. For me the idea of closure as reproduction of the organization of the system is perfectly acceptable. It seems like a wonderful insight. But I don't share the feeling Maturana and Varela have that organization is a discrete state. According to them, if a system's organization changes, the system is no longer the same system—it is a different system if its organization changes. It seems to me that organization exists, on the contrary, on a continuum and not as a discrete state. Consider, for example, evolution, in which all kinds of small innovations and mutational possibilities are

tried out in different environments. It's problematic when these mutational possibilities constitute a new species. Drawing distinctions between
species is to some extent arbitrary, especially when there is an extensive
fossil record. There are many instances in contemporary ecologies where
it is impossible to say if an organism falls within the same species or
constitutes a different species. Clearly the organizational pattern of that
system has changed in a substantive way, enough to allow one to make a
distinction, but the change falls along a spectrum. It is not black and
white—either no change, or a completely different system. While it's an
important insight to see that the living is intimately bound up with the
reproduction of a system's organization, I don't see that it's necessary to
insist there is a definitive closure in what constitutes an organization.

cw: The way I read Maturana and Varela's point is in a more cognitive or
epistemological register, which is to say that if you observe something,
you either call it X or not-X, X or Y, and that to cognize *at all* is to engage
in the making of that distinction. Your description, it seems to me, is
talking as if all these things are going on out there in nature, and then the
question is, do our representations match up with them or not? That
seems to me to be the pretty strongly realist and representationalist
premise of the scenario you just described.

kh: Yes, but in this I don't differ in the least from Maturana and Varela,
who are constantly using arguments based on exactly the kind of natural
history case studies that I just mentioned in order to demonstrate the
closure of the organism. I grant your point, that I'm assuming there is
some way to gain reliable knowledge about these things. And of course
it's always possible to open up scientific "facts"—or as Bruno Latour calls
them, "black boxes"—and bring them into question again. But one has to
argue from some basis.

wr: Can I ask you, Professor Luhmann, about your black box? In a sense,
your black box is operational closure, beyond which you will not go. You
do not want to dispense with it; it's the fundamental element of your system or your theory. As we discussed before, if we are talking about leaps
of faith, that's your leap of faith. What is at stake in retaining operational
closure? Why is it so important for your theory?

nl: Certainly, in sociological theory, or in social theory in general, you
have the problem of how to distinguish objects or areas of, say, law, the
economy, and so on. You can say that the economy is essentially coping
with scarcity, or something like this. And to avoid these kinds of essentialist

assumptions, I try to say that the law is what the law says it is, and the economy is just what the economy in its own operation produces out of itself. This is, I think, the alternative, in which I try to opt for a tautological definition. And then I'm obliged to characterize how the operations of the system—say, communication as the characteristic operation of society—follow a certain binary code, like legal versus illegal, to be able to reproduce, say, the legal system. Recursive decision making reproduces an organization. But then I have this problem: I do not share the opinion of Maturana and Varela that outside relations *are* cognition, that you have already a cognitive theory if you say "operational closure." Maturana and Varela present structural coupling, structural drift, and these terms as cognitive terms. But I would rather think that a system is always, in its operation, beyond any possible cognition, and it has to follow up its own activity, to look at it in retrospect, to make sense out of what has already happened, to make sense out of what *was* already produced as a difference between system and environment. So first the system produces a difference of system and environment, and then it learns to control its own body and not the environment to make a difference in the system. So cognition then becomes a secondary achievement in a sense, tied to a specific operation which, I think, is that of making a distinction and indicating one side and not the other. It's an explosion of possibilities, if you always have the whole world present in your distinctions.

wr: OK, maybe we should move on to the topic of negation. Could you summarize for us, Professor Hayles, your use of the semiotic square in your notion of double negation in your article on constrained constructivism?

kh: I don't know how to give a short answer to this, so I'll have to give the long answer.

wr: Good.

kh: As I understand Greimas's work, he developed the semiotic square in order to make simple binaries reveal complexities that are always encoded in them but that are repressed through the action of the binary dualism. The idea is to start with the binary dualism and then, by working out certain formal relationships, to make it reveal implications that the operation of the binary suppresses. To give you an example, consider Nancy Leys Stepan's article about the relation between race and gender in physiognomic studies in the late nineteenth and early twentieth century. Stepan notes the circulation within the culture of expressions like "women are the blacks of Europe." To analyze this expresession, consider a semiotic square that begins with the duality "men and women." What implications are present in that duality which aren't fully explicit? Some of those implications can be revealed by putting it in conjunction with another duality, white/black. By using the semiotic square and expanding the men/women duality, it is possible to demonstrate, as Ronald Schleifer and his coauthors have done, that "men" really means "(white) men" and "white" really means "white (men)." By developing formal relationships of the semiotic square, one can make the duality yield up its implications. It is important to remember that there is no unique solution to a semiotic square. Any duality will have many implications encoded into it, connotations which are enfolded into that duality but which are not formally acknowledged in it. So there are many sets of other dualities that can be put in conjunction with the primary one. If they're doing the work they should do in a semiotic square, each second pair would reveal different sets of implications. This is a preface to explain what I think the semiotic square is designed to do. Beyond this, the semiotic square is formally precise. It is Greimas's hypothesis that there are certain formal relationships that dictate how dualities develop. So it's not arbitrary how the relationships within the square are developed.

In the semiotic square I used, I wanted a binary which is associated with scientific realism: true and false. If a hypothesis is congruent with the world it's true. Popper argued that science cannot prove truth, only falsity. According to him, a hypothesis must be falsifiable to be considered scientific. The true/false binary is rooted in scientific realism. In order to have the "true" category occupied, you have to be in some objective, transcendent position from which you can look at reality as it is. Then you can match your hypotheses up with the world and see if the two are congruent. Thus the true/false binary comes directly out of realist assumptions. The binary I proposed to complicate and unravel the true/false dichotomy with was "not-false" and "not-true." I claim that the "true" position cannot be occupied because there is no transcendent position from which to say a hypothesis is congruent with reality. The "false" position *can* be occupied, because hypotheses can be falsified, as Popper argued. More ambiguous is the "not-false" position. This position implies that within the realm of representations we construct, a hypothesis is not inconsistent with the unmediated flux. Notice it is not true, only consistent with our interactions with the flux. Even more ambiguous is the "not-true" position; it represents the realm of possibilities which have not been tested, which have not even perhaps been formulated, and which may never be formulated because they may lie outside the spectrum of realizable experiences for that species. It is this position on the lower left of the square, the negative of a negative, that is more fecund, for it is the least specified and hence the most productive of new insights. Hence Shoshana Felman's phrase for it, "elusive negativity."

CW: It's very important to you, it seems to me, to insist that those other possibilities that are opened up are not solely possibilities dependent upon the context of inquiry. This goes back to what you were talking about earlier with the unmediated flux containing or acting as a constraint, a finite set of possibilities—that's what these constraints finally are. So it's important to you to insist, versus say Maturana, that these unfolding possibilities do not tell us only about the *context* of inquiry, but about the object of inquiry. Would that be fair to say?

KH: Yes. That would be true to say.

WR: What is your reaction to the schema?

NL: Well, again, a long one. The first is that I would distinguish between making a distinction and positive/negative coding, so that negation comes into my theory only by the creation of language, and with the special purpose of avoiding the teleological structure of communication, its tendency to go by itself to a fixed position, to a fixed point, to a consensus point. So, if we have a situation in which every communication can be answered by "yes" or by "no"—I accept or I reject your proposal—then every selection opens again into either conformity or conflict. So, negation in this sense comes into my theory of society only by coding language, or doubling language so to speak, in a "yes" version and a "no" version. And of course it is important that you have the identity of the reference, the possibility to say "yes" or "no" to the same thing, and not to something else. I say "this is a banana," and you can say yes or no, but if you think that maybe it is an apple, then you have to make a distinction to talk about this. So this concerns negation. But I have also, independently of this, thought about an open question concerning distinction: distinction from what? And there are in principle, I think, two possibilities: distinction of an object from an unmarked space, from everything else (again, this is not a glass of wine, and not a tree, and so on). So, one type of distinction is that you create an unmarked space by picking out something. But then there is another type of distinction where you can cross the boundaries—male/female, for example, or in this example, true/false. And then you can oscillate between the two, and say, well, this is a job for a man or a job for a woman, is this good or is this bad, is this expensive, given our budget, and so on. But if you *can* indicate both sides by this distinction, then you also create by this very distinction an unmarked space, because then you can change from the distinction true/false to the distinction good/bad. Or to the distinction male/female. And then you can make a kind of correlation or coupling between different distinctions. But this always creates the world, creates an unmarked space, a kind of thing which you cannot indicate. Or if you indicate the unmarked space, then you have two marks, marked and unmarked.

wr: Then you'll have another unmarked space . . .

nl: Yeah, yeah, then you create another unmarked space beyond this distinction. And if I look at this fourfold scheme of Greimas's, I think that first it is quite clear that false/true is a specifiable distinction, specifiable on both sides. You can give arguments for true and you can give arguments for false, and you can have true arguments that something is false and false arguments that something is true. In this sense, it is complete. But then, when you make this distinction you also specify the unity of this distinction—which is, I would say, the code of science—and then you do not use, say, a political code (powerful or less powerful), or the gender code, or the moral code, or the legal code, or the economic code, or whatever. And when I look at this enlargement, I wonder whether it would be possible to say that indeed the false/true distinction is *not* a complete description of the world, that it leaves out the unmarked space, or it leaves out what you do not imagine, what you do not see, what you do not indicate, if you operate within this kind of framework. And this is important for my theory of functional differentiation, because if I identify codes and systems, then of course I need always a third value or third position: the rejection of all other codes. So, if I am in the legal code, then I am not in the economic code; the judge doesn't make his decision according to what he is paid for his decision . . .

cw: Sometimes! [laughter]

nl: Well, yes, but then that's a problem of functional differentiation. And if I look at Greimas's table with its four positions, I think first that the lower line, the "not-true" and "not-false" line, is simply representing the unmarked space. Then I would change the positions; in other words, I would make the distinction between "false" and "not-false." "False" is something which is verified as "false"; "not-false" is everything else. Or "true" and "not-true." I don't know whether this makes any sense, but the essential point is that for my theory, especially for the theory of functional differentiation, we need something which Gotthard Guenther would call "transjunctional operation"—that means going from a positive/negative distinction to a metadistinction, rejecting or accepting this kind of distinction. And you can, of course, have a metadistinction, then a meta-metadistinction, and that would always mean "marked/unmarked." And at that point, of course, you are in the middle of the question of how systems evolve by marking, by making marks in an unmarked space, and then you can have a history of possible correlations between structural developments and semantic developments in the history of society.

EK: Now your reinterpretation of this scheme, Professor Hayles, makes it look like it can no longer fulfill the function that, as I understand you, it's supposed to fill: namely, as far as I understand it, it's supposed to somehow assure us that we can somehow reach out of language and get language into contact with some sort of physical constraint. And when you interpreted the scheme ...

WR: Negation is simply part of ...

EK: ... part of the inside. Then you don't need a constraint anymore. I mean ...

NL: ... self-imposed constraint ...

EK: ... in your reinterpretation of the scheme you get rid of the external constraints, and I think I have trouble really understanding how we can reach, with the square, the idea of an external constraint.

WR: I guess the question is, how? What evidence does double negativity give? What evidence not only of the outside world, but in a sense what evidence does double negativity give that it does deal with ...

KH: It does not give any evidence, I think. I did not intend to say that it gave evidence. But Professor Luhmann was, I think, exactly right in identifying something in that second line with what he calls the unmarked, that which lies outside distinction, and that's exactly the category that I meant to designate by "not-true." "Not-true" is absence of truth, which is not to say that it's untrue; it's to say that it is beyond the realm in which one can make judgments of truth and falsity. It's an undistinguished area in which that distinction does not operate. So his idea of distinctions is very applicable to what I was trying to do there. What I was trying to ask was, is there a place in language that points toward our ability to connect with the unmediated flux? This does not prove that the unmediated flux exists; it does not prove that the unmediated flux is consistent; it does not prove that the unmediated flux operates itself through constraints. It's simply asking the question, if we posit the unmediated flux, then where is the place in language that points toward that connection? That place is "not-true" or "elusive negativity," because that's the area in language itself which points towards the possibility I'm trying to articulate as "unmediated flux." It's no accident, I think, that in Greimas's article on the semiotic square he talks about this position emerging through the constraints that are present in the structure of language itself. In other words, his idea is that the structure of the semiotic square is not arbitrary; it's embedded in the deep structure of language. That, of course, is a debatable proposition. But just say for a moment that we accept the proposition. Then my argument is that the structural possibilities offered to us by language contain logically and semantically a category which points toward something we cannot grasp but is already encoded into our language.

cw: Can I jump in here at this specific point? What I hear you saying is that language as such does not presuppose any particular referent, but it does presuppose reference as such, right? Would that be fair?

kh: Well, I don't know that I was saying that. I thought I was saying that language has a logical structure, and part of that logical structure is to provide for a space for the unknowable and the unspeakable, even though paradoxically that space has to be provided within the linguistic domain.

cw: Right, but it's presupposed that it could be knowable and could be speakable, and moreover that that knowable and speakable is finite, right?

kh: The knowable and the speakable . . .

cw: . . . or contains a finite set of applications in language.

kh: What is in the category "absence of truth" could always be brought into the category of either "not-false" or "false." It would be possible to have a scientific theory which brings something which was previously unthought and unrecognized into an area of falsifiability. But no matter how much is brought into the area of falsifiability, it does not exhaust and cannot exhaust the repertoire of those possibilities. So, this goes back to Professor Luhmann's idea that there is a complexity outside systems which is always richer than any distinction can possibly articulate.

cw: I guess the difference that I'm trying to locate here is that, in Professor Luhmann's scheme, this outer space is automatically produced by the deployment of distinctions—marking produces an unmarked space—but the difference is, in principle it seems to me, your claim about constraint, as we talked about it earlier: that it depends upon this set being finite. For you, it's not possible in principle to just go on and on and on deploying yet another distinction.

KH: Right.

CW: Because otherwise the claims about reality and the constraints that it imposes seem to me to fall apart at that point.

KH: Well, here maybe I can invoke some ideas about mathematics and say that I'm not sure the range of things that can be brought in to the realm of "not-false" and "false" is finite. It may be infinite, but if it is infinite, then it is a smaller infinity than the infinity of the unmediated flux, and as you know, Cantor proved the idea that one infinity can be smaller than another. So, if it's an infinity, it is a smaller infinity than the set of all possibilities of all possible constructions.

NL: In my terms, you would then have the question, what do you exclude as unmarked if you make the distinction between infinite and finite? [Laughter] But that's a book of Philip Herbst from the Tavistock Institute entitled *Alternatives to Hierarchies,* where he refers to Spencer Brown and raises somewhere the question, what is the primary distinction? You could have the distinction finite/infinite, you could have the distinction inside/outside, you could have the distinction being/not-being to start with, and then you can develop all kinds of distinctions in a more or less ontological framework. And I find this fascinating, that there is no exclusive, one right beginning for making a distinction. The classics would of course say "being" and "not-being," and then the romantics would say infinite/finite, and systems theory would say inside/outside. But how are these related? If you engage in one primary distinction, then how do the others come again into your theory or not? This is part of the postmodern idea that there is no right beginning, no beginning in the sense that you *have* to make one certain distinction and you can fully describe the start of your operations. And that's the background against which I always ask, "What is the unity of a distinction?" Or "What you do exclude if you use this distinction and not another one?"

CW: For me at least, the interest of your work, both of you, is that it is trying to take that next step beyond the mere staging or positing of incommensurable discourses. It seems to me that both of you—in finally somewhat opposed ways—are trying to move beyond this paradigmatic type of postmodern thought and move on—in your case, Professor Luhmann—to what you call a universally applicable or valid description of social systems. And in your case, Professor Hayles, that effort is revealed in your attempt to work out this problem of constraints—in a way, to try to rescue some sort of representationalist framework—to say that in fact there is a reality out there that does pose constraints and, moreover, can be known in different and specifiable ways by these discourses. It's possible, in other words, to see beyond that incommensurability . . .

KH: Yes, though I would not say—this sounds like a nit-picking correction, but to me it's the essence of what I'm trying to say—I wouldn't say that what is out there can be known; I would say our interaction with what is out there can be known.

CW: Then I think the question has to be, for me at least, in what sense are you using the term "objectivity" at the end of the "Constrained Constructivism" essay? A point that Maturana makes in one of his essays is that to use the subjective/objective distinction is to automatically presuppose or fall back on representationalist notions, which immediately recasts the problem in terms of realism and idealism.

KH: I don't use the word "objectivity."

CW: I have the *New Orleans Review* version . . .

KH: I don't think I use it in that essay . . .

CW: "In the process,"—this is about three paragraphs from the end . . .

KH: . . . oh, OK . . .

CW: ". . . in the process, objectivity of any kind has gotten a bad name. I think this is a mistake, for the possibility of distinguishing a theory consistent with reality, and one that is not, can also be liberating"—and you go on to talk about how this might be enabling politically, which is, I think, interesting because it does accept the challenge of moving beyond just saying, "well it's all incommensurable."

KH: Here, I accept the kinds of arguments that have been made by Donna Haraway and Sandra Harding about "strong objectivity," that to pretend one does not have a position is in fact not being "objective," in the privileged sense of "objective," because it ignores all those factors that are determining what one sees. And to acknowledge one's positionality, and explore the relationship between the components that go into making up that position and what one sees, in fact begins to allow one to see how those two are interrelated, and therefore to envision other possibilities. Sandra Harding's formulation of "strong objectivity" takes positionality into account, and is therefore a stronger version of objectivity than an objectivity that is based on some kind of transcendent nonposition.

cw: Let me follow up here. I guess the problem I have, and this is the case with Harding's work, is that what you're describing is inclusion. I see how that means more democratic representation of different points of view, but I don't see how it adds up to "objectivity" in the sense that it's usually used. Unless the sense of objectivity here is procedural, that we all agree to follow certain rules of a given discourse.

kh: As a philosopher, Harding doesn't want to relinquish the term "objectivity."

cw: Yes, that's quite clear.

kh: I don't have any vested interest in keeping the word "objectivity," but I think the idea of what she's pointing to, whether one calls it "objectivity" or not, is no matter how many positions you have, they will not add up to a transcendent nonlocation.

cw: Right. The God's-eye view.

kh: P_1 plus P_2 plus ... P_n is not God.

cw: Right.

wr: So actually what you're talking about is what you mentioned in the very beginning: the word "objectivity" basically means "reflexivity"—the reflexivity that you were missing in the early cybernetic tradition?

kh: Yes. I don't know if anybody's used the word "strong reflexivity," but I would like to. Strong reflexivity shows how one can use one's position to extend one's knowledge. That's part of what is implied in the idea that we know the world because we are connected to it. Our connection to it is precisely our position. Acknowledging that position and exploring precisely what the connections are between the particularities of that position and the formations of knowledge that we generate, is a way to extend knowledge. There is a version of reflexivity that, in the early period of science studies, was like an admission of guilt: "Well, I'm a white male, and so therefore I think this." There was a period when you couldn't write an article without including a brief autobiography on who you were. But that really missed the point, because the idea is to explore in a systematic way what these correlations are, and precisely why they lead to certain knowledge formations, and therefore to begin to get a sense of what is not seen.

NL: Then my opponent should be not so much for the term "objectivity," but for the term "interaction," and who sees the interaction.

WR: Interaction between us and an environment . . .

NL: Yeah, yeah. I have no trouble in posing external observers, a sociologist who sees an interaction between the capitalistic economy and the political system, or between underdeveloped countries—center/periphery, and so on—but how could we think that the system that interacts with its environment is itself observing the interaction as something which gives a more or less representational view of what is outside? How can we see this without seeing that this is a system which does the observing? How could we avoid involving the system—which means a radically constructivist point of view—when we ask the question, "who is the observer?" We say "the outside observer, of course." He sees interactions of any kind, causal or whatever, as objective reality in *his* environment, because *he* sees it. But if the system *in* interaction tries to *see* the interaction, how could we conceive this?

KH: There may be many ways to use the word "interaction," and I'm not sure I'm using it in the sense you mean. For me, when I say the word "interaction," it already presupposes a place prior to observation, whether self-observation or observation by someone else. It's the ultimate point that we can push to in imagination, it's the boundary between the perceptual apparatus and the unmediated flux, and as such it is anterior to and prior to any possible observation. So, I would say that the interaction is not observable.

NL: Then you can drop the concept.

KH: You could drop the concept, except then you have a completely different system. What interaction preserves that I think is important is the sense of regularities in the world and the guiding role that the world plays in our perception of it. If representation and naive realism, with their focus on external reality, only played one side of the street, Maturana's theory of autopoiesis, with its focus on the interior organization of systems, only plays the other side. I am interested in what happens at the dividing line, where one side meets the other side. Maturana's theory is important for me because it shows, forcefully and lucidly, how important perception and systemic organization are in accounting for our view of the world. It also opens the door to a much deeper use of reflexivity than had been possible before—an insight significantly extended by your positioning of the observer as he (or she) who makes the distinctions that bring systems into existence as such. But for me, this is not the whole picture. If it is true that "reality is what we do not see when we see," then it is also true that "our interaction with reality is what we see when we see." That interaction has two, not one, components—what we bring to it, and what the unmediated flux brings to it. The regularities that comprise scientific "laws" do not originate solely in our perception; they also have a basis in our interactions with reality. Omitting the zone of interaction cuts out the very connectedness to the world that for me is at the center of understanding scientific epistemology.

WR: Well, I think that we've hit that outer limit right here, where we are redefining boundaries. Do we have any other general questions? Maybe the system in question ought to be dinner . . .

CW: Let me just ask one more very general question, since we're on this point, and it's something we've touched on. At the end of the "Constrained Constructivism" article, Professor Hayles, you make it clear that this rethinking you're engaged in has pretty direct ethical imperatives. Objectivity, for you and for Sandra Harding and for Donna Haraway, is an ethical imperative as well as an epistemological or theoretical one, and you go on to specify what those imperatives are. I take it for you, Professor Luhmann, that you want to be very careful to separate ethics as just one of many social systems from other types of social systems, all of which can be described by systems theory. So what I'm wondering is, could you all talk a little bit about what you see as the ethical and political imperatives, if there are any, of second-order theory, to reach back to where we started.

KH: I don't know that I really have anything to add beyond what you just said, but it is clear for me that there are ethical implications of strong reflexivity and strong objectivity. I'm not really versed in ethics as a kind of formal system, so I'll defer that to Professor Luhmann.

NL: Well, for me ethics or morality is a special type of distinction, and a particularly dangerous one, because you engage in making judgments about others—they are good or bad. And then if you don't have consensus, you have to look for better means to convince them or to force them to agree. There is a very old European tradition of this, the relation between standards and discrimination. If somebody is not on your side, then he is on the wrong side. And I think my work is a sociologist's way to simply reflect on what we engage in if we use ethical terms as a primary distinction in justifying our cognitive results: if you accept this you are good, and if you don't, you have to justify yourself.

Works Cited

Brown, George Spencer. *Laws of Form*. 2d ed. Reprint, New York: Dutton, 1979.

Greimas, A. J. "The Interaction of Semiotic Constraints." In *On Meaning: Selected Writings in Semiotic Theory*. Translated by Paul J. Perron and Frank H. Collins. Minneapolis: University of Minnesota Press, 1987.

Guenther, Gotthard. "Life as Poly-Contexturality." In *Beiträge zur Grundlegung einer operationsfähigen Dialektik*. Vol. 2. Hamburg: Meiner, 1979.

Haraway, Donna J. "Situated Knowledges: The Science Question in Feminism as a Site of Discourse on the Privilege of Partial Perspective." *Feminist Studies* 14 (1988): 575-99.

Harding, Sandra. "Eurocentric Scientific Illiteracy—A Challenge for the World Community." Introduction to *The "Racial" Economy of Science*. Bloomington: Indiana University Press, 1994.

Hayles, N. Katherine. "Constrained Constructivism: Locating Scientific Inquiry in the Theater of Representation." *New Orleans Review* 18, no. 2 (spring 1991): 76–85.

Heims, Steve Joshua. *Constructing a Social Science for Postwar America: The Cybernetics Group, 1946–1953*. Cambridge: MIT Press, 1993.

Herbst, P. G. *Alternatives to Hierarchies*. Leiden: Nijhoff, 1976.

Letvin, J. Y., et al. "What the Frog's Eye Tells the Frog's Brain." *Proceedings of the Institute for Radio Engineers* 47 (1959): 1940–51.

Luhmann, Niklas. "The Cognitive Program of Constructivism and a Reality That Remains Unknown." In *Selforganization: Portrait of a Scientific Revolution*, edited by Wolfgang Krohn, Gunter Kuppers, and Helga Nowotny. Dordrecht: Kluwer Academic, 1990.

Lyotard, Jean-François. *The Differend: Phrases in Dispute*. Translated by Georges Van Den Abbeele. Minneapolis: University of Minnesota Press, 1988.

Maturana, Humberto. "Science and Daily Life: The Ontology of Scientific Explanation." In *Research and Reflexivity*, edited by Frederick Steier. London: Sage, 1991.

Maturana, Humberto, and Francisco Varela. *Autopoiesis and Cognition: The Realization of the Living.* Dordrecht and Boston: D. Reidel, 1985.

————. *The Tree of Knowledge: The Biological Roots of Human Understanding.* Rev. ed. Translated by Robert Paolucci. Boston: Shambhala, 1992.

Popper, Karl L. *Conjectures and Refutations: The Growth of Scientific Knowledge.* 2d ed. New York: Basic Books, 1965.

Stepan, Nancy Leys. "Race and Gender: The Role of Analogy in Science." *Isis* 77 (1986): 261–77.

Varela, Francisco, Evan Thompson, and Eleanor Rosch. *The Embodied Mind: Cognitive Science and Human Understanding.* Cambridge: MIT Press, 1993.

von Foerster, Heinz. *Observing Systems.* Seaside, Calif.: Intersystems, 1981.

7. Making the Cut: The Interplay of Narrative and System, or What Systems Theory Can't See

N. Katherine Hayles

The originary moment for the creation of a system, according to Niklas Luhmann, comes when an observer makes a cut ("Cognitive Program"). Before the cut—before any cut—is made, only an undifferentiated complexity exists, impossible to comprehend in its noisy multifariousness. Imagine a child at the moment of birth, assaulted by a cacophony of noise, light, smells, and pressures, with few if any distinctions to guide her through this riot of information. The cut helps to tame the noise of the world by introducing a distinction, which can be understood in its elemental sense as a form, a boundary between inside and outside (Brown). What is inside is further divided and organized as other distinctions flow from this first distinction, exfoliating and expanding, distinction on distinction, until a full-fledged system is in place. What is outside is left behind, an undifferentiated unity. Other cuts can be made upon it, of course, generating other systems. But no matter how many cuts are made there will always be an excess, an area of undifferentiation that can be understood only as the other side of the cut, the outside of the form.

It is no accident that this story has a mythopoetic quality, for it is a mythology as much as a description. It is a way of explaining how systems come into existence that perform two tasks at once: it describes the generation of systems, and it also constructs the world as it appears from the viewpoint of systems theory. As the story indicates, the primary distinction necessary to be able to think systems theory is a cut that divides system from environment. According to systems theory's own account, however, there is also an outside to this cut, an area that from the viewpoint of systems theory can be seen only as a mass of undifferentiated world tissue. Another way to organize this material, I suggest, is narrative.

The coexistence of narrative with system can be seen in Luhmann's account of the creation of a system, for his account is, of course, itself a narrative. Its very presence suggests that systems theory needs narrative as a supplement, just as much, perhaps, as narrative needs at least an implicit system to generate itself. Narrative reveals what systems theory occludes; systems theory articulates what narrative struggles to see.

In constructing a narrative that will contest systems theory's account of how meaning is generated, I will follow Luhmann's advice. To get beyond the space enclosed by a system's assumptions, he recommends looking at ideas that, within the confines of a given system, can appear only as paradoxes or contradictions. One enlarges or escapes from a system, he believes, by interrogating what cannot be made logical, straight, or ordered within the system. As we know, his version of systems theory begins with an observer making a distinction. Where does this observer come from? Is he brought into view through the action of another observer looking at him? If so, where does this second observer come from? The problem is not solved by supposing that the observer observes himself, for then we must ask where this capacity to observe himself comes from. If we pose the question logically, as systems theory would have us do, it cannot be answered within the system, for it leads only to an infinite regress of observers, each of whom is constituted in turn by another observer.

Suppose we take another path and construct the question as a historical inquiry. From what intellectual predecessor, what preexisting body of discourse, does Luhmann draw in order to think of beginning with the observer? The answer to that question is clear, for Luhmann himself provides it. This way of thinking about systems comes from a modification of autopoiesis, a concept defined and developed by the noted Chilean neurophysiologist, Humberto Maturana. To get outside systems theory and interrogate what it cannot see, I will begin with a historical and narrative account of Maturana's work. More is at issue in this interrogation than Luhmann's construction of systems and Maturana's epistemology, influential as they are. I seek to understand the tension between narrative and systemic thinking in general. Why does Foucault, especially in his early work, have such difficulty accounting for epistemic shifts? Why does Lacan's account of psychological formations insist that women can find no way to represent themselves? Why does any system, once it is exposed by a systems theorist, tend to seem inescapable and coercive? To get a purchase on these questions, let us look at systems theory from the other side of the cut, that is, from narrative rather than the proliferating distinctions that constitute systems.

Cutting Away the World: Defining the Living as a Closed System

Maturana's epistemology is grounded in studies of perception. In the famous article "What the Frog's Eye Tells the Frog's Brain," Maturana and his coauthors demonstrated that a frog's visual system operates very differently from that of a human (Lettvin, Maturana, McCulloch, and Pitts). Small objects in fast, erratic motion elicit maximum response, while large, slow-moving objects evoke little or no response. It is easy to see how such perceptual equipment could be adaptive from a frog's point of view, because it allows him to perceive flies while ignoring other phenomena irrelevant to his interests. The results imply that the frog's perceptual system does not so much register reality as construct it. As the authors put it, their work "shows that the [frog's] eye speaks to the brain in a language already highly organized and interpreted instead of transmitting some more or less accurate copy of the distribution of light upon the receptors" (1950). The work led Maturana to the maxim fundamental to his epistemology: "Everything said is said by an observer" (Maturana and Varela, *Autopoiesis and Cognition*, xxii).

Despite the potentially radical implications of the article's content, however, its form reinscribed the conventional realist assumptions of scientific discourse. Nowhere do the authors acknowledge that the reality they report is constructed by their sensory equipment no less than the frog's is by his. Faced with this inconsistency, Maturana had a choice. He could continue to work within the prevailing assumptions of scientific objectivity, or he could devise a new epistemology that would construct a picture of the world consistent with what he thought the experimental work showed. The break came with his work on color vision in primates, specifically humans. He and his coauthors found that that they could not map the visible world of color upon the activity of the nervous system (Maturana, Uribe, and Frenk). There was no one-to-one correlation between perception and the world. They could, however, correlate activity in a subject's retina with his color experience. If we think of sense receptors as constituting a boundary between outside and inside, this result implies that organizationally the retina matches up with the inside, not the outside. From this and other studies, Maturana concluded that perception is not fundamentally representational. As Maturana recounts in *Autopoiesis and Cognition,* he and his coauthors decided to treat "the activity of the nervous system as determined by the nervous system itself, and not by the external world; thus the external world would have only a triggering role in the release of the internally-determined activity of the nervous system" (xv).

Maturana's key insight was to realize that if the action of the nervous system is determined by its own organization, the result is necessarily a circular, self-reflexive dynamic. The organization of a system is constituted through the processes it engages in, and the processes it engages in are determined by its organization. To describe this circularity, he coined the term *autopoiesis,* or self-making. "It is the circularity of its organization that makes a living system a unit of interactions," he and Varela wrote in *Autopoiesis and Cognition,* "and it is this circularity that it must maintain in order to remain a living system and to retain its identity through different interactions" (9). He regarded the autopoietic closure of the space a system inhabits as the necessary and sufficient condition for it to be alive. Building on this premise of autopoietic closure, Maturana developed a new and startlingly different account of how we know the world.

Here let me pause for a digression. Before discussing Maturana's epistemology, I want to register an objection to the leap he makes when he goes from saying perception is nonrepresentational to claiming it has no connection with the external world. In my view, his data do not justify this larger claim. Other researchers, among them Walter Freeman and Christine Skarda, have also argued against a representational model of perception (Skarda). Freeman and Skarda's data on the olfactory perception of rabbits are akin to Maturana's results, in that the data indicate the rabbit's responses are transformative and highly nonlinear, influenced not only by the experience at hand but also by previous experiences the animal has had, his emotional state at the moment, and a host of other factors. To say the relation is transformative is different, however, from claiming there is no relation. The divorce of perception from external reality is at once the basis for the striking originality of Maturana's epistemology, and the Achilles heel that renders it vulnerable to cogent objections.

What is this epistemology? I will approach it in an anecdotal and narrative fashion, a rhetorical mode quite different from the highly abstract and reflexive language of *Autopoiesis and Cognition,* the landmark work Maturana coauthored with Francisco Varela. (Later I will have more to say about the mode of Maturana's exposition and the purposes his rhetorical formulations serve.) To enter Maturana's world, consider how the world would look from the point of view of one of your internal organs, say your liver. To imagine this fully, you will need to leave behind as much of your anthropomorphic orientation as possible. Your liver has no plans for the future or regrets about the past; for it, past and future do not exist. There is only the present and the ongoing processes in which it engages. Similarly, since your liver has no way to conjoin cause and effect, causality does not exist for it. If you drink excessive amounts of alcohol, it may develop cirrhosis, for it is structurally coupled to its environment and its processes change in coordination with changes in the environment. This coupling does not, however, constitute causality. The causal link you discern between drinking and cirrhosis is constructed by you as an observer; your liver knows nothing of it.

Maturana's denial of causality is worth exploring in more depth. It is at once counterintuitive and central to his epistemology. Consistent with his premise of operational closure, he maintains that no information is exchanged between a system and its environment. Events that happen in the environment do not cause anything to occur in the living organism. Rather, they are the historical occasions for triggering actions determined by a system's organization. The difference between an event "triggering" an action and "causing" it may seem to be a quibble, but for Maturana, the distinction is crucial. Causality implies that information moves across the boundary separating an organism from its environment and that it makes something happen on the other side. Say you slap me and I become angry. In the conventional view, one would say that your slap caused me to be angry. As this inference indicates, a causal viewpoint organizes the world into subject and object, mover and moved, transmitter and receiver. The world of causality is also the world of domination and control. Maturana sought to undo this perception by positing that living systems are operationally closed with respect to information. A system acts always and only in accord with its organization. Thus events can trigger actions, but they cannot cause them because the nature and form of a system's actions are self-determined by its organization. For example, if I am a masochist, I may be pleased rather than angry at your slap. Your slap is only the historical occasion for the self-determined processes that I engage in as a result of being structurally coupled to my environment.

One implication of letting go of causality is that systems always behave as they should, which is to say, they always operate in accord with their structure, whatever that may be. In Maturana's world, my car always works. It is I as an observer who decides that my car is not working because it will not start. Such "punctuations," as Maturana calls them, belong to the "domain of the observer" (*Autopoiesis and Cognition* 55–56). Because they are extrinsic to the autopoietic processes, they are also extrinsic to the biological description that Maturana aims to give of life and cognition. To accommodate the difference in states between, say, a car that will and will not start, Maturana makes a sharp distinction between structure and organization. Structure refers to the actual state of a system at a given moment. Structure changes over time as an organism grows, ages, contracts disease, recovers health. Organization, by contrast, defines the nature of the organism as such. Organization can be thought of as the complete repertoire of all the structures that the organism can exhibit and still remain that organism. When a system's organization changes, it ceases to be that kind of system and becomes something else, for example dead rather than alive. Always leery of reification, Maturana stressed that organization, as a concept, exists only in the domain of the observer. On the level of autopoietic process, it is not a concept but an instantiated reality implicit in the constitutive relations of the processes to each other.

It should be apparent by now that the cut Maturana makes between the observer and autopoietic process is intended to act as a prophylactic barrier against anthropomorphism. Our commonsense intuitions about the world are relegated to the "domain of the observer," leaving the space of autopoiesis free from contamination by time, causality, motivation, intentionality, and desire. Thus emptied, the autopoietic space feels surprisingly serene, in much the same way that Buddhist notions of emptiness are serene. (It is interesting in this regard that Varela, Maturana's coauthor, later connected his own version of embodied cognition with Buddhist philosophy [Varela, Rosch, and Thompson].) But serenity comes at a price. Autopoiesis, in the case of conscious organisms, must contain the observer, yet the observer, with his anthropomorphic projections and causal inferences, is precisely what has to be excluded for autopoiesis to come into view as such. The strain of these contradictory necessities can be seen in Maturana's construction of cognition. Clearly cognition must emerge from autopoietic processes if it is not to be treated as an ad hoc phenomenon, a soul injected into the machine. But what kind of cognition can autopoiesis produce? Because Maturana wants to eradicate anthropomorphic projections from his account of the living, the cognition that he sees bubbling up from autopoiesis is empty of representational content. It can thus scarcely qualify as conscious thought. At most it precedes or underlies the familiar lifeworld of representation that we occupy (or that occupies us).

The divorce of consciousness from autopoietic process results in a curious gap in the theory's circular reasoning. How do we know autopoietic processes exist? Recall that Maturana's epistemology is grounded in perceptual studies of the frog's visual system and primate color vision, among others. According to his epistemology, these studies (along with every other construction that presupposes time, causality, and representation) rely on concepts that are not intrinsic to autopoiesis but rather are punctuations introduced by observers. We know autopoietic processes exist because of these studies, but the epistemology of autopoiesis requires that these studies be regarded as "punctuations" extraneous and irrelevant to autopoiesis. The circularity that is one of the theory's strongest and most striking features is here interrupted by the cordon separating the observer from the processes that must nevertheless somehow give rise to her.

The quarantine of the observer also requires that Maturana ignore the feedback loops that connect the observer with her autopoietic processes. Suppose I have stalled my car on the railroad tracks and, as I struggle to get it started, I see a train speeding toward me. The future moment when the train will strike the car exists vividly in my imagination, and I have no difficulty foreseeing the causal chain of events that will splatter me and my car over the landscape. As a result of these punctuations, which according to Maturana exist only in my domain as an observer, my heartbeat accelerates, my respiration alters dramatically, and my endocrine system releases a flood of adrenaline into my body. Evidently, the observer is not only an observer but also an intrinsic part of the autopoietic totality. Why does this story, or its analogue, never get told in *Autopoiesis and Cognition*? To answer this question, I must take my narrative onto new ground and consider the rhetorical strategies that *Autopoiesis and Cognition* uses to construct its argument. How the story is told is also part of the story.

Self-Making as Literary Form: The Rhetoric of Autopoiesis

Aside from the introductions, *Autopoiesis and Cognition* consists of two essays, "Biology of Cognition" and "Autopoiesis: The Organization of the Living." In both essays, the writing is almost exclusively analytic, with one proposition related to another logically in an argument that proceeds by division and subdivision, implication and extension. There are only two examples of narrative, and they stand out because they are so unusual. In one, the authors illustrate the difference between an ordinary and an autopoietic viewpoint by supposing that two teams of builders are put to work constructing a house (53–55). The first team is told it is building a house, and each craftsman understands his work in that context. With the second team, no mention is made of a house. Rather, each craftsman is given a copy of a set of instructions and told which parts he is supposed to execute. In both instances the finished product is the same—the house is built. The first team thinks that it has been building a house all along, however, whereas the second team thinks only that it has been engaging in a set of specified processes. The authors use the example to illustrate how a seemingly teleological project can emerge from processes that have no awareness of a larger goal. Although they do not interpret the example this way, it can also be used to illustrate why narrative is in tension with autopoiesis. The set of processes that the second team actuates could not be a story, or rather not a story anyone would find interesting, for lacking any sense of purpose, causality, or goal, it would consist only of a series of statements such as "This is happening, and then this is happening, and then something else is happening." To be effective, narrative requires a sense of how the present relates to past and future and of at least potentially causal relations between events.

The second anecdote is even more revealing. In it, the authors imagine that a man is piloting a plane by following his instrument readings (50–51). When he lands, his friends and family congratulate him on his excellent feat of navigation. He is amazed at their admiration, for from his point of view, he has only been manipulating the controls so that the dials on his instruments stay within specified limits. Repeated with slight variations several times in Maturana's writing (sometimes the pilot is in a plane, at other times in a submarine), this anecdote evidently has special meaning for him. Tyrone Cashman speculates on its significance in an imaginary dialogue he constructs between Maturana and Sartre. He impishly has Sartre suggest that Maturana's epistemology, like Sartre's own views, were influenced by childhood experiences, particularly Maturana's poor eyesight. Sartre recalls a joke Maturana likes to tell on himself about being so nearsighted as a child that he could not tell the difference, until his brother pointed it out to him, between a stout lady waiting for a bus and a mailbox. Sartre says it is no wonder that Maturana makes a cut separating the observer from autopoietic processes, for what he observed as a child was indeed a punctuation different from what was there in reality. In Maturana's theory, Sartre observes, the world as we know it comes into existence when it is constructed by two observers "languaging" between themselves. Maturana's epistemology thus reinscribes the linguistic acts of distinction that took place when his brother told him that the heavyset woman was not, after all, a repository for mail. How would this epistemology hold up, Sartre wonders, for "a rural child with sharp eyesight, who before the age of ten spent a great deal of time alone, by himself or herself, exploring woodlands and streams and lake shores, observing insects and the stages of plant life, stalking wild animals and listening to the subtle changes of bird calls—to such a person, your theory might sound absurd. Languaging, for him or her, precisely inhibits good observation. When someone else is present, the natural world is perceived less vividly and richly" (Cashman 6–7).

This story, which we can consider a counternarrative to Maturana's anecdote about the pilot who flies blind, illustrates one of the dangers of narrative for someone who wants to construct a system. Unlike analytic writing, narrative is contextual. Instead of relying on numbered subdivisions to advance its plot, as Maturana's analytic writing does, narrative uses description. Inherent in the contextualization of narrative is a certain "loose bagginess" (as Henry James called it), for example, language necessary to set a scene or move the story from one locale to another. In Maturana's anecdote, there are phrases that put the man into the plane (or submarine) and take him out of it, even though these actions are not relevant to the story's point—relevant, that is, in his interpretation. As Cashman's send-up makes clear, what is extraneous and irrelevant in one reading can become highly relevant in another. Because narrative is contextual, it is polysemous in a way that analytic writing is not. Getting a narrative to mean only one thing is like getting a bowl of wiggling Jell-O to have only one shape. The medium won't allow it.

In addition to its contextuality, narrative differs from analytic writing in its use of historical contingency. When Maturana uses numbers to move from one statement to another, he is employing a semiotic system whose order is not in doubt, thus implying that the relation between his numbered statements is as definitive and noncontingent as the progression one, two, three. Narrative, by contrast, characteristically reinscribes historical contingency, relating events that might have happened other than they did. It was not inevitable that Maturana would be extremely nearsighted as a child and not wear corrective lenses; nor was it foreordained that Sartre as a child would be left alone to spend long days in the woods. Things just happened this way and (in Cashman's interpretation) later bore fruit in the two competing epistemologies. In contrast to these historical contingencies are the logical necessities that Maturana seeks to reveal through his analysis. Frequently, when he is obliged by custom or literary form to comment on his analytic writing (as in the introduction to *Autopoiesis and Cognition* or the "Comments" section of a journal article) he will express impatience, claiming that the piece is complete in itself and that to add anything further would be extraneous. These comments suggest that he regards his analytic writing as constituting a kind of closed autopoietic space in itself, secure in its circular organization and insulated against historical contingency. To bring that assertion (or illusion) of closure into question, I turn now to an account of the historical contingencies that connect Maturana's theory to its predecessors in the Macy conferences.

Accidents of History: How Homeostasis Became Autopoiesis

As is often the case with heuristic examples, Maturana's anecdote about the navigator did not come out of nowhere. It had a predecessor in the Macy conferences. Funded by the Josiah Macy Foundation, the Macy conferences were annual affairs and ran for nearly a decade, from 1946 to 1953. Attendance was by invitation only. The idea was to bring together a group of researchers working at the forefront of their fields to forge a new interdisciplinary paradigm that became known, retrospectively, as cybernetics. Christened by Norbert Wiener, cybernetics was conceived as a science that would develop a common explanatory framework to talk about animals, machines, and humans by considering them as information processors that encoded and decoded messages, exacerbated or corrected their actions through feedback loops, and demonstrated circular causality. (See Heims for an account of the Macy conferences.)

A key concept in the Macy conferences was homeostasis. Understood as the ability of a system to maintain stability by keeping its parameters within certain limits, homeostasis was discussed in a context that made clear its relation to World War II. If homeostasis failed, W. Ross Ashby pointed out, the result was death, whereas if it succeeded, "your life would be safe" (von Foerster 79). Ashby illustrated the concept with an anecdote about an engineer in a submarine. The engineer avoids catastrophe by keeping the ship's parameters stable. As a biological organism, he is a homeostatic system in a feedback loop with the ship, which is also a homeostatic system; he keeps its homeostasis functioning, and as a result, he can maintain his own homeostasis as well. The example alludes to a situation that, in the context of the recent war, was resonant with danger; the man's vulnerable situation metonymically stood for the larger peril of a society drawn back from the brink of destruction. In the wake of the war homeostasis had a strongly positive valence, for it was the scientific counterpart to the "return to normalcy" that the larger society was fervently trying to accomplish.

To illustrate homeostasis, Ashby constructed an electrical device he called a homeostat that operated with transducers and variable resistors. When it received an input changing its state, the homeostat searched for the configuration of variables that would return it to its initial condition. In the postwar context, it seemed obvious that homeostatic calculations must include the environment. If the environment is radically unstable, the individual organism cannot continue to survive. "Our question is how the organism is going to struggle with its environment," Ashby remarked, "and if that question is to be treated adequately, we must assume some specific environment (von Foerster 73–74). This specificity was expressed through the homeostat's four units, which could be arranged in various configurations to simulate organism-plus-environment. For example, one unit could be designated "organism" and the remaining three the "environment"; in another arrangement, three of the units might be the "organism," with the remaining one the "environment." Ashby arranged the mechanism so that if the homeostat did not compensate for environmental changes within specified limits, it overloaded or "died."

Elsewhere I have suggested that in the Macy conferences, homeostasis became the nucleus for a cluster of concepts that emphasized equilibrium and stability (Hayles). The homeostasis constellation developed in relation and opposition to another constellation centered on reflexivity. Through the idea of the feedback loop, homeostasis already had built into it the notion of circular causality. The man in the submarine, when he manipulates the dials, effects a change in some variable, say the air pressure in the control room. As a result, the oxygen level increases, and the man can think more clearly and operate the dials more efficiently. Thus the causal chain he initiated circles around to include his system as a biological organism as well. Applied to language, circular causality opened up a passage into the dangerous and convoluted territory of reflexivity, for it implied that an utterance is at once a statement about the outside world and a reflection of the person who uttered it.

It is significant that the word "reflexivity" does not occur in the Macy transcripts. Although the participants were struggling with ideas that, in contemporary usage, are commonly associated with reflexivity, the lack of a central term meant that the discussion was often diffuse, spreading out into diverse metaphors and discursive registers. The most intense debate about what I am calling reflexivity was embedded in a discourse that had its own assumptions, only one of which was reflexivity. This discourse was psychoanalysis. The conjunction between reflexivity and psychoanalysis was forged in the presentations made by Lawrence Kubie, a Freudian psychoanalyst associated with the Yale University Psychiatric Clinic. By all accounts, Kubie was a tendentious personality. Certainly his presentations evoked strong resistance from many participants, especially the physical scientists. As if to demonstrate circular causality, his repeated attempts to convince the scientists of the validity of psychoanalytic theory became more intransigent as they met resistance, and they evoked more resistance as they became more intransigent. Kubie's central message was that language is always multiply encoded, revealing more than the speaker realizes. When some of the scientists objected to this idea, wondering what evidence supported it, Kubie in personal correspondence interpreted their resistance as hostility that itself required psychoanalytic interpretation. It is no wonder that the scientists were enraged, for in Kubie's hands, language became a tar baby that stuck to them the more they tried to push it away. The association of reflexivity with psychoanalysis meant, for many of the participants, that the concept was a dead end that had little or no scientific usefulness. Not only could it not be quantified, it also subverted normative assumptions about scientific objectivity.

The particularities of this situation—Kubie's halitosis of the personality, the embedding of reflexivity within psychoanalytic discourse, the unquantifiability of the concepts as Kubie presented them—put a spin on reflexivity that affected its subsequent development. The people at the Macy conferences who were convinced that reflexivity was a crucially important concept (including Margaret Mead, Gregory Bateson, and Heinz von Foerster) were marked by the objections it met within that context (see Brand for anecdotal evidence to this effect). The influence of these historical contingencies can be seen in von Foerster's treatment of reflexivity in *Observing Systems*. The punning title announces reflexivity as a central theme. "Observing" is what (human) systems do; in another sense, (human) systems themselves can be observed. The earliest essay in the collection, taken from a presentation given in 1960, shows that von Foerster was thinking about reflexivity as a kind of circular dynamic that could be used to solve the problem of solipsism. How does he know that other people exist, he asks? Because he experiences them in his imagination. His experience leads him to believe that they similarly experience him in their imaginations. "If I assume that I am the sole reality, it turns out that I am the imagination of somebody else, who in turn assumes that he is the sole reality" (7). In a circle of intersecting solipsisms, the subject uses his imagination to conceive of someone else, and then of the imagination of that person, in which he finds himself reflected; and so he is reassured not only of the other person's existence, but of his own as well. That even a fledgling philosopher could reduce this argument to shreds is perhaps beside the point. Von Foerster seems to recognize that it is the philosophical equivalent to pulling a rabbit from a hat, for he purports to "solve" the paradox by asserting what he was to prove, namely the existence of reality.

Although the argument is far from rigorous, it is interesting for the line of thought it suggests. Even more revealing is the cartoon (drawn at his request by Gordon Pask) of a man in a bowler hat, in whose head is pictured another man in a bowler hat, in whose head is yet another man in a bowler hat. The potentially infinite regress of men in bowler hats does more than create an image of the observer who observes himself by observing another. It also bears a striking resemblance to Maturana's phrase "domain of the observer," for it visually isolates the observer as a discrete system inside the larger system of the organism as a whole. The correspondence is not accidental. In the aftermath of the Macy conferences, one of the central problems with reflexivity was how to talk about it without falling into solipsism or resorting to psychoanalysis. The message from the Macy conferences was clear: If reflexivity was to be credible, it had to be insulated against subjectivity and presented in a context where it had at least the potential for rigorous (preferably mathematical) formulation. As Norbert Wiener was later to proclaim, "Cybernetics is mathematics or it is nothing" (Wiener).

Throughout the 1960s, von Foerster remained convinced of the importance of reflexivity and experimented with various ways to formulate it. A breakthrough occurred in 1969, when he invited Maturana to speak at a conference at the University of Illinois. Maturana used the occasion to unveil his theory of "cognition as a biological phenomenon" (*Autopoiesis and Cognition* xvi). The power of Maturana's theory must have deeply affected von Foerster, for his thinking about reflexivity takes a quantum leap up in complexity after this date. In his 1970 essay "Molecular Ethology: An Immodest Proposal for Semantic Clarification," he critiques behaviorism by making the characteristically reflexive move of turning the focus from the observation back onto the observer. Behaviorism does not demonstrate that animals are black boxes that give predictable outputs for given inputs, he argues. Rather, it shows the cleverness and power of the experimenter in getting them to behave as such. "Instead of searching for mechanisms in the environment that turn organisms into trivial machines, we have to find the mechanisms within the organisms that enable them to turn their environment into a trivial machine" (von Foerster 171).

By 1972, the influence of Maturana on von Foerster is unmistakable. In his 1972 essay "Notes on an Epistemology for Living Things," he casts Maturana's theory of autopoiesis into numbered quasi-mathematical propositions and gives it a circular structure, with the last proposition referring the reader back to the beginning (von Foerster). The influence was mutual, for von Foerster's idea that the observer is located in an isolated arena became incorporated into Maturana's theory. Recall that von Foerster produced the observer through imagining an infinite regress of men in bowler hats; something of this ad hoc production lingers in Maturana's conception. If we ask where Maturana's observer comes from, it is apparent that he is not a biological production, which would imply a physiological explanation of how autopoiesis gives rise to consciousness. (The absence of such explanation is scarcely surprising, given that contemporary cognitive science lacked a detailed picture of how consciousness bubbles up from autopoietic processes.) Rather, the production of the observer is accomplished rhetorically, by positing an enclosed space called "the domain of the observer." Not coincidentally, the enclosure of the observer in this domain also creates a sanitized space where reflexivity can be acknowledged without rebounding back to ensnare the observer in every utterance he makes. In fact, just the opposite happens. The observations of the observer reflect back on himself but do not have efficacy in explaining autopoietic processes, which happen on their own in another sphere that is constructed to be objective precisely because it excludes the observer from its informationally closed space. Reflexivity is thus rehabilitated from the taint of subjectivity it received from its association with psychoanalysis in the Macy discussions, but at the cost of erecting a prophylactic barrier between the observer and autopoietic processes.

Here it may be useful for me to pause and reflect, in reflexive fashion, on the kind of argument I have been fashioning. Whereas the systems approach Maturana uses presents his theory as an autonomous entity sufficient in itself, the narrative approach I have been following shows how Maturana's theory both drew on and changed the concepts that preceded it. These changes did not happen gratuitously. At least in part, they were in response to particular historical contexts that had invested the constellations of homeostasis and reflexivity with specific interpretations, values, and problematics. What logic is to system, historical contingency is to narrative. Had Kubie had a different personality, or had von Foerster not constructed the observer in terms of solipsism, or had Maturana not been invited to the Chicago conference, the reflexivity constellation might have developed other than it did. While narrative may reach toward something approaching inevitability in seeing events as multiply determined, the kind of closure it evokes is qualitatively different from that which emerges from systems theory. The inevitabilities derive not from logical necessity but from contingency piled on contingency.

Also different are the continuities narrative traces between what came before and what happened after. Whereas the systems approach treats systems as self-contained unities, the narrative approach sees systems coming into existence through patterns of overlapping replication and innovation. New ideas are woven not out of whole cloth (even cloth must have its precedessors in thread, loom, and pattern) but are forged out of previous instantiations and contexts that are partly changed and partly replicated. The term I appropriated (from archeological anthropology) to describe this pattern of overlapping replication and innovation is seriation. To see seriation in action, consider what happens to homeostasis as a concept evolving in specific historical contexts. As we have seen, for Ashby and his colleagues, homeostasis included the system plus the environment. Moreover, it used circular causality—that is, feedback loops between the system and environment—to return the system to equilibrium. The homeostat was an instantiation of a goal-seeking machine whose goal was stability. When it achieved stability, it was successful and "lived"; when it lurched into instability, it failed and "died." Considered essential for survival, homeostasis was thus linked to the idea of the living organism, although it included mechanical (and more speculatively, social) systems as well. In this respect, it carried out the imperative of the cybernetic program to create a common framework for animals, humans, and machines.

When Maturana took it over, he redefined homeostasis so that the circle of causality no longer went from the system to the environment but rather was contained internally within the autopoietic processes. At the same time, he manifested his allegiance to biology by leaving behind mechanical and social systems and making the closure of the autopoietic space the necessary and sufficient condition for a system to be living. He kept the idea of a goal, but now the goal was the continued production of the autopoietic space rather than stability. The goal of autopoiesis is more autopoiesis. Stability remained linked with survival, but the entities that were to be kept stable were redefined. No longer did survival demand that state variables had to remain within certain limits, as with homeostasis. Rather, the crucial entity that had to remain stable to ensure survival was organization. Instantiated within the autopoietic processes, a system's organization must persist unchanged through time for the system to retain its identity as such.

By showing seriation at work, I do not mean to imply that autopoiesis, as a theory, is defective or patched up. In fact, seriation usually works in the opposite direction of progressive refinement and fuller realization of the new elements that have entered the picture. Nor is it a reflection on

Maturana's originality to show that he appropriated ideas from models that preceded his. According to my argument, almost everyone does. Indeed, Maturana's theory is striking in its boldness and in its uncompromising vision of moving beyond anthropomorphic concepts of life. Although autopoiesis emerged from homeostasis, it is also substantially different from it, as I have indicated above. It actually represents a blending of ideas from both the homeostasis and reflexivity constellations. From homeostasis it appropriated stability, endurance, and survival; from reflexivity, the circular structure of a system turning back on itself to create a closed, self-referential space. It also explicitly rejects ideas that, in the Macy conferences, are associated both with reflexivity and homeostasis, for example circular causality (recall that for Maturana, causality does not exist in itself but only as a connection made in the domain of the observer).

The innovations that make autopoiesis different from homeostasis are clearly laid out by Paul Dell, a family systems theorist who has been at the forefront of the movement to apply autopoiesis to the field of family therapy. Dell points out several ways in which the language of homeostasis contains implications that are incompatible with autopoiesis. The one most relevant here, perhaps, is his argument that whereas homeostasis implies that a system will remain the same, autopoiesis implies a system will change. When Ashby designed the homeostat, he conceptualized it as a mechanism that searched for a function E-1 that compensates for a function E expressing complex change in the environment. As a result of this compensation, the machine's variables remain within specified limits. Its purpose, on this view, is to return the system back to an equilibrium whenever it is disturbed. From an autopoietic viewpoint, by contrast, the system is a system precisely in the sense that its components interact with each other; none can be separated out from the whole. Moreover, the system never reacts to changes in the environment, only to changes within itself triggered by its structural coupling with the environment. If one component changes—if, for example, the daughter of an alcoholic father ceases to facilitate his drinking—all of the other components have to change as well, because the interactions between them have changed. This reasoning implies that from an autopoietic viewpoint, change anywhere in the system drives the system toward a new configuration rather than back toward a prior equilibrium point.

Put this way, autopoiesis sounds as if it ought to be amenable to narrative progression, despite the self-circularity of its theoretical structure. The idea is put to the test in another book Maturana coauthored with

Varela. *The Tree of Knowledge* proposes to articulate autopoiesis together with the theory of evolution. Because the theory of evolution is about change and historical contingency, it is fundamentally narrative. I have been suggesting that systems theory and narrative constitute opposite approaches to the construction of meaning. What happens when systems theory meets evolution?

The Circle Versus the Line: A Disjointed Articulation

The circular structure of autopoiesis provides the inspiration for the literary form of *The Tree of Knowledge*. As the opening diagram of the chapters indicates, the authors envision each chapter leading into the next, with the final one coming back to the beginning. "We shall follow a rigorous conceptual itinerary," they announce in the introduction, "wherein every concept builds on preceding ones, until the whole is an indissociable network" (9). The structure is meant to enact their central idea that "all doing is knowing and all knowing is doing" (27) by showing the interrelation between simple and complex living systems. Accordingly, they start with unicellular organisms (first-order systems), progress to multicellular organisms with nervous systems (second-order systems), and finally to cognitively aware humans who interact through language (third-order systems). Humans are made up of cells, of course, so cellular mechanisms must be at work in complex systems as well; in this way, the end connects with the beginning. Autopoiesis, the continuing production of processes that produce themselves, is the governing idea connecting systems at all levels, from the single cell to the most complex thinking being. "What defines [living systems] is their autopoietic organization, and it is in this autopoietic organization that they become real and specify themselves at the same time" (48). Instantiating a linear narrative that turns into a circle, the book simulates an autopoietic structure in which the details produce the overall organization, and the organization produces the details. Traversing this path, the "doing" of the reader—the linear turning of pages as she reads—becomes also a kind of "knowing," for she experiences the structure of autopoiesis as well as comprehends it when the text circles back on itself.

The problem comes when the authors try to articulate this circular structure together with evolutionary "lineages"—literally, the creation of lines. In evolution, lineage carries both the sense of continuity (traced far enough back, all life originates in single-cell organisms) and qualitative change, as different lines branch off from one another and follow separate evolutionary pathways. Here I want to mark an important difference between evolution and autopoiesis: whereas in autopoiesis lines become circles, in evolution lines proliferate into more lines as speciation takes place through such mechanisms as genetic diversity and differential rates of reproductive success. In an attempt to finesse this difference, Maturana and Varela proclaim repeatedly that for an organism to continue living, it must conserve autopoiesis as well as adaptation. And how does it do this? By remaining structurally coupled to its environment. As incremental changes occur in the environment, corresponding incremental changes also occur in the organism. Thus the organism always remains within the circle of autopoiesis, but this circular motion can also move along a line, as when a rolling ball falls downhill. "Ongoing structural change of living beings with conservation of their autopoiesis is occurring at every moment, continuously, in many ways at the same time. It is the throbbing of all life" (100).

The articulation of autopoiesis with evolution thus hinges on the claim that structures gradually evolve while still conserving autopoiesis. To describe the change that takes place, the authors use the term "natural drift." There seems to be a natural drift in "natural drift," however, and in later passages it becomes "structural drift." If structure changes, what does it mean to say that autopoiesis is conserved? Here they fall back on the distinction between structure and organization they had previously used in *Autopoiesis and Cognition*: "Organization denotes those relations that must exist among the components of a system for it to be a member of a specific class. Structure denotes the components and relations that actually constitute a particular unity and make its organization real" (47). Interestingly, they use a mechanical rather than a biological analogy to illustrate the distinction. A toilet's parts can be made of wood or plastic; these different materials correspond to differences in structure. Regardless of the material used, however, it will still be a toilet if it has a toilet's organization (47). The analogy is strangely inappropriate for biology. For life forms based on protein replication, it is not the material that changes but the way the material is organized.

What does it mean, then, to claim that autopoiesis is conserved? According to them, it means that organization is conserved. And what is organization? "Those relations that must exist among the components of a system for it to be a member of a specific class" (47). These definitions force one to choose between two horns of a dilemma. Consider the case of an amoeba and a human. Either an amoeba and a human have the same organization, which would make them members of the same class, in which case evolutionary lineages disappear because every living system has the same organization; or else an amoeba and a human have different organizations, in which case organization—and hence autopoiesis—must not have been conserved somewhere (or many places) along the line. The dilemma reveals the tension between the conservative circularity of autopoiesis and the linear thrust of evolution. Either organization is conserved and evolutionary change is effaced, or organization changes and autopoiesis is effaced. Contrary to the authors' assertions, the circle cannot be seamlessly articulated with the line. Whatever recuperations the authors attempt through their title, the tree Darwin used to image descent has a branching structure that remains at odds with the circularity of autopoiesis.

The strain of trying to articulate autopoiesis with evolution is most apparent, perhaps, in what is not said. Genetics is scarcely mentioned, and then in contexts that underplay its importance. At one point, the authors acknowledge that "modern studies in genetics have centered mainly on the genetics of nuclear acids," but they suggest that other heredity systems have been obscured by this emphasis, including "those associated with other cellular components such as mitochondria and membranes" (69). Elsewhere they acknowledge that they have "skimmed over" populations genetics but claim that "it is not necessary to scrutinize the underlying mechanisms" (i.e., genetics) to understand "the basic features of the phenomenon of historical transformation of living beings" (115). In the absence of any discussion of genetics, how do they explain evolutionary change? Through an organism's structural coupling with the environment, combined with the structural diversity introduced by [sexual] reproduction. One is left with the impression that the primary mechanism of evolution is structural change within an organism due to its interactions with its environment, which are passed on to its offspring. "To sum up: evolution is a natural drift, a product of the conservation of autopoiesis and adaptation" (117). Thus they concur with Lamarck and Darwin, placing themselves outside the synthesis between evolution and genetics that produced contemporary evolutionary biology.

Given their emphasis on autopoiesis, it is perhaps obvious why they choose to sidestep genetics, for any discussion of genetics would immediately make clear that the distinction between structure and organization cannot be absolute—and if this distinction goes, then autopoiesis is no longer conserved in evolutionary processes. For if organization is construed to mean the biological classes characterized as species, then it is apparent that organization changes as speciation takes place. If organization means something other than species, then it ceases to distinguish between different kinds of species and simply becomes instead the property of any living system. Conserving organization means conserving life, which may be adequate for autopoiesis to qualify as a property of living systems but does nothing to articulate autopoiesis with evolutionary change.

The essential problem here is not primarily one of definitions, although it becomes manifest at these sites in the text because they are used to anchor the argument, which otherwise drifts off into such nebulous terms as "natural drift." Rather, the difficulties arise because of Maturana's passionate desire to have something conserved in the midst of continuous and often dramatic change. Leaving aside the hand-waving explanations of structure and organization, that something is basically the integrity of a self-contained, self-perpetuating system that is operationally closed to its environment. In Maturana's metaphysics, the system closes on itself and leaves historical contingency on the outside. Even when he is concerned with the linear branching structures of evolution, he turns this linearity into a circle and tries to invest it with a sense of inevitability. Narrative is encapsulated within system, like a fly within amber. Seen as a textual technology, *The Tree of Knowledge* is an engine of knowledge production that vaporizes contingency by continuously circulating within the space of its interlocking assumptions.

Like many postwar systems, including Foucault's epistemes and Lacan's psycholinguistics, autopoiesis is profoundly subversive of individual agency. It therefore makes an interesting comparison with Richard Dawkins's idea of the "selfish gene," another theory that locates the essence of life in aconscious processes rather than conscious subjectivity. Whereas Maturana elides genetics, Dawkins foregrounds it. This difference reflects a deeper divergence in their treatments of agency. Dawkins images humans as "lumbering robots" controlled by their genes, but agency is not missing from his scheme; it is simply displaced from the conscious mind into the genes. The social and economic formations associated with rampant individualism, especially capitalism, are as vigorous as ever in Dawkins's rhetoric and narratives. The players may have shrunk to microscopic size, but the rules of the game—and the stakes it entails—remain the same. Maturana, by contrast, constructs agency as a contest over how the boundaries are drawn that constitute systems. Complex systems are made up of parts that are themselves autopoietic entities. Thus a human is constituted through its cells, which in turn are made up of yet smaller entities. Which of these autopoietic systems is subordinate to which? The answer, for Maturana, is not so much a given

as an ethical imperative that depends on prior assumptions about freedom and what he unashamedly calls love. In an organism, the component unities are properly subordinated to the organismic whole. The case
is different for a society. There, the system exists for the benefit of its
component parts, namely individual humans. For Maturana, autopoiesis
resides finally and most forcefully at the level of the individual.

Other than as an ethical imperative, why this should be so remains
shrouded in mystery. Because past and future do not exist in Maturana's
scheme except as modes of existing in the present, it is not possible to
ground this imperative in a myth of origin. "The business of living keeps
no records concerning origins," he and Varela write in *The Tree of Knowledge.* "All we can do is generate explanations, through language, that reveal the mechanism of bringing forth a world. By existing, we generate
cognitive 'blind spots' that can be cleared only through generating new
blind spots in another domain. We do not see what we do not see, and
what we do not see does not exist" (242). One of these cognitive blind
spots, I have been arguing, is narrative. And one of the windows that
opens onto it, I have further suggested, is the construction of the observer
in systems theory. When Niklas Luhmann makes the move of turning
the construction of the observer into an origin, he departs from the circularity of autopoiesis and begins a new cycle in the seriated pattern of overlapping innovation and replication that lies at the heart of my narrative.

The Observer as Origin: Luhmann's Reinscription of Maturana

When Luhmann begins with the observer, in a stroke he does away with
the difficulties Maturana encountered by rhetorically constituting the
observer within a separate "domain" isolated from the autopoietic system
("Cognitive Program"; *Essays on Self-Reference*). Far from being impris-
oned within the system and existing in an ad hoc relation to it, the observer now generates the system by drawing a distinction. The reflexivity
that appeared so threatening within the context of the Macy conferences,
after being sanitized and encapsulated in an isolated domain, thus reenters
the system at a foundational moment.

Perhaps Luhmann felt free to make this move because he is primarily
concerned with social theory rather than biology. As a social theorist, he
obviously does not have the same stake as Maturana in avoiding anthropomorphic projections of what life is.

Just as Maturana redescribed terms and shifted emphases when he in-
scribed into biology ideas appropriated from cybernetics, so Luhmann
changes as well as reinscribes autopoiesis when he takes it into social

theory. Insofar as Maturana succeeds in linking autopoiesis with life, he wins for it a central position within biology, for it addresses a concern fundamental to the discipline. When Luhmann applies autopoiesis to social systems, he is led by this history to say that social systems are alive. But in importing the claim into a different disciplinary context, he also changes its position. Whereas for Maturana the connection with life is crucial, in Luhmann the claim that social systems are alive does no interesting work within his theory and, indeed, is scarcely developed beyond this bare assertion. It rather exists as a skeuomorph, that is, a feature that served an instrumental purpose in previous instantiations but now works as an allusion and a link to the past. The fabric of seriation is woven out of skeuomorphs as much as innovations.

The pattern of seriation can also be seen in Luhmann's appropriation of other ideas central to autopoiesis. Consider Maturana's postulate that autopoietic systems are informationally closed and that they always conserve their organization. In Luhmann these ideas are transformed into his premise that social systems are operationally closed (*Differentiation of Society*; "Operational Closure"). The difference between operational and informational closure is revealing. Recall that for Maturana, the idea of closure was grounded in his studies of perception. For Luhmann perception is more or less beside the point, since he is dealing with societies rather than organisms. Accordingly, the mechanism of closure is displaced from the working of perception onto the working of codes. One system cannot communicate with another because they employ different codes; the operations that a system can perform are defined and contained by the codes it employs. The circularity of autopoiesis is thus realized for Luhmann in the interplay between a system's codes and its organization. The operations it performs through its codes define its organization, and its organization defines the codes. In Maturana, the essence of life is displaced from (human) consciousness onto aconscious autopoietic processes. In Luhmann, this displacement is registered as the play of codes within a system. Luhmann does not see social interactions as exchanges between purposeful individuals with complex psychologies. Rather, interaction takes place between the codes that social agents employ. It is the codes, not the agent's conscious thoughts or perceptions, that structure the situation. When one goes out to drink, one employs the code of drinking, and it is this code, not the individual's thoughts or activities, that constitute drinking as drinking. What autopoietic biological processes are to Maturana, social codes are to Luhmann.

We saw earlier that Maturana constructed agency in terms of where a system's boundaries are drawn. From societies to organisms to cells, systems are complex unities that themselves are composed of systems that are complex unities. The question of which level can or should subordinate the other levels to itself is for Maturana an ethical issue that cannot be decided within systems theory. There is nothing inherent in the nature of systems that dictates the organism should dominate its subsystems, whereas a society should be subordinate to its subsystems. Maturana recognizes the fragility of this argument when he identifies it with love. In a theory remarkable for the circularity of its interlocking premises, love enters as excess, emerging not from necessity but desire. A similar dynamic is played out in Luhmann's work in his idea of a functionally differentiated society. Whereas in medieval times societies were organized vertically, with each subsystem subordinate to the larger system that contained it, in the modern period some societies have achieved a horizontal structure that enables different subsystems to operate independently of one another. This is the kind of structure that Luhmann prefers, for he believes it fosters diversity and minimizes coercion. But he recognizes that there is nothing inevitable about its emergence. Indeed, he regards it as sufficiently improbable so that it is at any time liable to collapse and revert back to a hierarchical structure, as happened in Nazi Germany. Thus the fragility of love and the vulnerability of desire is replicated as well as changed in Luhmann's reinscription of autopoiesis.

Of all these seriated relationships, perhaps the most crucial is the one with which we began: the point at which the observer is inserted into the system. By moving the observer to the point of entry or origin, Luhmann opens the system—any system—to alternative constructions. As a result, although his systems are no less closed than Maturana's, the activity of system making is considerably more open. The difference is registered in the phrase that Luhmann adapts from Maturana's dictum "we do not see what we do not see." In a reinscription that is also an innovation, Luhmann writes that "reality is what one does not perceive when one perceives it" ("Cognitive Program" 76). Like Maturana, Luhmann postulates a realm that remains apart from the constructed world of human perception. But unlike Maturana, he twists the closed circle of tautological repetition ("we do not see what we do not see") into an asymmetric figure ("one does not perceive when one perceives"). The energy generated by these contradictory propositions rebounds like a loaded spring toward the very term that Maturana's closure was designed to erase, namely "reality." What is enacted rhetorically within the structure of this sentence is formalized in Luhmann's theory by investing the observer with the agency to draw a distinction. By making a distinction, the observer reduces the unfathomable complexity of undifferentiated reality into something she can understand; by proliferating distinction on distinction, she begins to reproduce within this space of differentiation some of the complexity and diversity of a reality that remains forever outside (*Differentiation of Society,* "Operational Closure").

The importance of Luhmann's positioning of the observer has been recognized in different ways by a number of theorists writing in this volume. William Rasch concentrates on the siren call of the transcendent, which he sees Luhmann, along with his predecessor Wittgenstein, successfully resisting; Jonathan Elmer notes parallels between Luhmann's theory and Lacan's construction of the observer in the mirror stage; Cary Wolfe argues persuasively that Luhmann's importance for this cultural moment lies in the alternative he offers to the seemingly endless quarrels between constructivists and realists. My argument seeks to position itself at the very point where the observer comes into view at all. When Luhmann acknowledges that the observer, by drawing a different distinction, can generate a different kind of system, he opens a trapdoor out of the coerciveness of systems. But this is a limited kind of escape, for in Luhmann's metaphysics, escape from one system is achieved only by entering another system. My efforts have been directed toward providing an alternative—not another system, but another way of organizing the material that is narrative rather than systemic.

To recapitulate: the advantage I claim for narrative is that it renders the closures that systems theory would perform contingent rather than inevitable, thus mitigating the coercive effects that systems theory can sometimes generate. As I see it, the problem with systems theory is that once a system stands revealed in all its pervasiveness and complexity— whether it be the invisible workings of power in Foucault's society of surveillance, or Lacan's psycholinguistics, or Maturana's autopoiesis—the system, precisely because of its logic and power, is likely to seem inevitable and inescapable. Among systems theorists, Luhmann is remarkable in seeing that every system has an outside that cannot be grasped from inside the system. If his own inclination is toward the closure of system rather than the contingency of narrative, he nevertheless has the intellectual honesty and generosity of spirit to see that closure, too, has an outside it cannot see. And this has given me room to argue that the very interlocking assumptions used to achieve closure are themselves the result of historical contingencies and embedded contextualities. Thus in my reading, a system looms not as an inevitability, but rather emerges as a historically specific construction that always could have been other than what it is, had the accidents of history been other than what they were. In this reading, one exits the system not merely to enter another system, but to explore the exhilarating and chaotic space of constructions that are contingent on time and place, dependent on specific women and men making situated decisions, partly building on what has gone before and partly reaching out toward the new.

Works Cited

Brand, Stewart. "'For God's Sake, Margaret': Conversation with Gregory Bateson and Margaret Mead." *The CoEvolution Quarterly* (summer 1976): 32–44.

Brown, George Spencer. *Laws of Form.* New York: Julian Press, 1972.

Cashman, Tyrone. "The Elysian Dialogs." *A Newsletter of Ideas in Cybernetics* 17 (summer 1989): 1–9.

Dawkins, Richard. *The Selfish Gene.* Oxford: Oxford University Press, 1976.

Dell, Paul F. "Beyond Homeostasis: Toward a Concept of Coherence." *Family Process* 21 (March 1982): 21–44.

Hayles, N. Katherine. "Boundary Work: Homeostasis, Reflexivity, and the Foundations of Cybernetics." *Configurations* 2 (fall 1994): 441–68.

Heims, Steve J. *The Cybernetics Group.* Cambridge: MIT Press, 1991.

Lettvin, J. Y., H. R. Maturana, W. S. McCulloch, and W. H. Pitts. "What the Frog's Eye Tells the Frog's Brain." *Proceedings of the Institute for Radio Engineers* 47 (1959): 1940–51.

Luhmann, Niklas. "The Cognitive Program of Constructivism and a Reality That Remains Unknown." In *Selforganization: Portrait of a Scientific Revolution,* edited by Wolfgang Krohn, Gunter Kuppers, and Helga Nowotny. Dordrecht: Kluwer, 1990.

———. *The Differentiation of Society.* New York: Columbia University Press, 1982.

———. *Essays on Self-Reference.* New York: Columbia University Press, 1990.

———. "Operational Closure and Structural Coupling: The Differentiation of the Legal System." 13 *Cardozo Law Review* 1419 (1992).

Maturana, Humberto R., G. Uribe, and S. Frenk. "A Biological Theory of Relativistic Color Coding in the Primate Retina." In *Archivos de Biología y Medicina Experimentales.* Suplemento 1. Santiago, Chile: 1968.

Maturana, Humberto R., and Francisco J. Varela. *Autopoiesis and Cognition: The Realization of the Living.* Dordrecht and Boston: D. Reidel, 1980.

———. *The Tree of Knowledge: The Biological Roots of Human Understanding.* Boston and London: New Science Library, 1987.

Skarda, Christine A. "Understanding Perception: Self-Organizing Neural Dynamics." *La Nuova Critica* 9–10 (1989): 49–60.

Varela, Francisco J., Eleanor Rosch, and Evan Thompson. *The Embodied Mind: Cognitive Science and Human Experience.* Cambridge: MIT Press, 1991.

von Foerster, Heinz. *Observing Systems.* Seaside, Calif.: Intersystems, 1981.

———, ed. *Cybernetics: Transactions of the Ninth Conference.* New York: The Josiah Macy Jr. Foundation, 1953.

Wiener, Norbert. *God and Golem, Inc.: A Comment on Certain Points Where Cybernetics Impinges on Religion.* Cambridge: MIT Press, 1964.

8. In Search of Posthumanist Theory: The Second-Order Cybernetics of Maturana and Varela

Cary Wolfe

In the current social and critical moment, no project is more overdue than the articulation of a posthumanist theoretical framework for a politics and ethics not grounded in the Enlightenment ideal of "Man." In what is called (for better or worse) postmodern theory, that humanist ideal is critiqued most forcefully, of course, by the early and middle Foucault, whose "genealogical" aim is to "account for the constitution of knowledges, discourses, domains of objects, etc., without having to make reference to a subject which is either transcendental in relation to the field of events or runs in empty sameness throughout the course of history" by virtue of his—and it must be "his"—privileged relation to *either* the presence or the absence of the phallus, language or the symbolic, property, productive capacity or tool making, and reason or a soul ("Truth and Power" 58). In Foucault, however, this call for posthumanist critique is more often than not accompanied, as many critics have noted, by a kind of dystopianism that imagines that the end of the humanist subject is the beginning of the total saturation of the social field by power, domination, and oppression.[1] And the later Foucault, as if compensating for his early dystopianism, evinces a kind of nostalgia for the Enlightenment humanism powerfully critiqued in his early and middle work but handled much more sympathetically in the *History of Sexuality* project.[2]

But posthumanist theory need not indulge either Foucauldian dystopianism or its compensatory nostalgia for the subject to critique the ethical and political separation of the human from the nonhuman on the basis of all that Bruno Latour has recently called the "magnificent features that the moderns have been able to depict and preserve": "the free agent, the citizen builder of the Leviathan, the distressing visage of the human person, the other of a relationship, consciousness, the *cogito,* the hermeneut, the inner self, the thee and thou of dialogue, presence to oneself, intersubjectivity" (136). As Latour recognizes, posthumanist theory cannot proceed simply by historicizing the human; instead, he argues, "we first have to relocate the human, to which humanism does not render sufficient justice" (136). And in this project of relocation, historical and dialectical means of situating the human are not enough.

Indeed, one need only think of the difficulties experienced by the Marxist tradition in theorizing the problem of ecology to see that the limitations of humanism and the legacy of the Enlightenment episteme are not solved by dialectical historicization alone.[3] Even within the Marxist tradition, a number of theorists have recognized that Marxism's liberation of "the total life of the individual" (to borrow Marx's phrase from *The German Ideology*) is purchased at the expense of its brutal objectification of nature and the nonhuman—a dynamic deeply symptomatic, in turn, of its Enlightenment inheritance, which imagines that man-the-producer liberates himself insofar as he fully exploits and raises himself above that object and resource called "nature." No one in the Marxist tradition recognized this more than Theodor Adorno, whose "negative dialectics" may be viewed (as critics as diverse as Fredric Jameson and Drucilla Cornell have suggested) as a kind of limit case in the attempt to "relocate" the human by historical, dialectical means. Adorno lamented that the historical dialectic of traditional Marxism would turn the whole of nature into "a giant workhouse" for an essentially imperialistic subject, and proposed instead "a thoroughgoing critique of identity" and "the Concept," which might enable thought to relocate the human in a field characterized by what Adorno called "the preponderance of the object"—of the nonidentity, heterogeneity, and multiplicity that is reduced and mastered by the identity term of the *positive* dialectic in its traditional form (*Negative Dialectics* 183).[4]

It is not enough, then, to hold onto the concept of human and simply embed it in networks of symbolic, discursive, and material production, for doing so would simply reenact the retreat and return of the subject-as-origin, which gave rise to Foucault's brilliant dismantling of this maneuver in essays like "Nietzsche, Genealogy, History." It means, rather, rethinking the notion of the human *tout court*—a project that fields outside of cultural and social theory have been vigorously engaged in over the past twenty years. In recent work in cognitive ethology, field ecology, cognitive science, and animal rights philosophy, for instance, it has become abundantly clear that the humanist habit—*especially* within "linguacentric" disciplines such as cultural criticism—of making even the possibility of subjectivity coterminous with the species barrier is deeply problematic.[5] This body of work has pursued the dismantling of humanism from a direction diametrically opposed to that of Foucault; instead of eroding the boundary between the human subject and its networks of production, it has taken the conceptualization of humanist subjectivity at its word and then shown how humanism must, if rigorously pursued, generate its own deconstruction once the traditional marks of the human (reason, language, tool use) are found beyond the species barrier. Donna Haraway summarizes many of these developments in her groundbreaking "A Cyborg Manifesto." "By the late twentieth century in United States scientific culture," she writes,

> the boundary between human and animal is thoroughly breached. The last beachheads of uniqueness have been polluted, if not turned into amusement parks—language, tool use, social behavior, mental events. Nothing really convincingly settles the separation of human and animal. . . . Movements for animal rights are not irrational denials of human uniqueness; they are clear-sighted recognition of connection across the discredited breach of nature and culture. (151–52)

It should not be assumed, however, that the ethical and political stakes in this boundary erosion are limited to the well-being of nonhuman animals alone. Indeed, the imperative of posthumanist critique may be seen from this vantage—and *is* seen by thinkers like Haraway—as of a piece with larger liberationist political projects that have historically had to battle against the strategic deployment of humanist discourse *against other human beings* for the purposes of oppression. Humanism, in other words, is species specific in its logic (which rigorously separates human from nonhuman) but not in its effects (such logic has historically been used to oppress both human and nonhuman others). As Gayatri Spivak points out in a recent essay,

> the great doctrines of identity of the ethical universal, in terms of which liberalism thought out its ethical programmes, played history false, because the identity was disengaged in terms of who was and who was not human. That's why all these projects, the justification of slavery, as well as the justification of Christianization, seemed to be alright; because, after all, these people had not graduated into humanhood, as it were. (229)

In this light, it is understandable that traditionally marginalized groups and peoples would be loath to surrender the idea of full humanist subjectivity, with all of its privileges, at just that historical moment that they seem poised to "graduate" into it. But as a host of theorists and critics of contemporary society have pointed out, it is not as if we have a choice about the coming of posthumanism; it is *already* upon us, most unmistakably in the sciences, technology, and medicine. Haraway has argued as forcefully as anyone that our current moment is irredeemably posthumanist because of the boundary breakdowns between animal and human, organism and machine, and the physical and nonphysical ("Manifesto," 151–55)—a triple hybridity we can find readily exemplified any evening on cable television, as in a recent program on the U. S. Navy's Marine Mammal project, in which highly trained bottlenose dolphins (human/animal) are fitted with video apparatuses (organism/machine) to locate underwater objects and beam their location back on the Cartesian grid of satellite mapping (physical/nonphysical).

For Haraway, the ethical and political implications of this sort of "cyborg" posthumanism are extremely ambivalent and totally inescapable. "From one perspective," she writes,

> a cyborg world is about the final imposition of a grid of control on the planet. . . . From another perspective, a cyborg world might be about lived social and bodily realities in which people are not afraid of their joint kinship with animals and machines, not afraid of permanently partial identities and contradictory standpoints. The political struggle is to see from both perspectives at once because each reveals both dominations and possibilities unimaginable from the other vantage point. (154)

Not surprisingly, then, the avoidance of the posthumanist imperative to "see from both perspectives" has, as Latour has recently pointed out, definite *pragmatic* ramifications. Humanist modernity, he argues in his recent study *We Have Never Been Modern,* is predicated upon a kind of paradox. On the one hand, modernity "creates mixtures between entirely new types of beings, hybrids of nature and culture." On the other, it "creates two entirely distinct ontological zones: that of human beings on the one hand; that of nonhumans on the other" (10–11). For Latour, this structure has the pragmatic payoff of enabling humanist modernity to "innovate on a large scale in the production of hybrids"—in the production, for example, of genetically engineered organisms like the aggressively marketed OncoMouse for cancer research—because the "absolute dichotomy between the order of Nature and that of Society" prevents the question of the dangerous mixture of ontological categories from ever arising (40, 42). But if the modernist constitutional separation of human and nonhuman has the practical advantage of allowing the proliferation of hybrid networks, it has the pragmatic drawback (as the strategy of re-pression always does) of ill-equipping contemporary society to explore in a thoughtful way how its relations to and in hybrid networks should be lived.

To do *that*, we must, Latour argues, move beyond the humanist con-stitution and rethink the very notion of politics itself: "The political task starts up again, at a new cost." He explains:

> It has been necessary to modify the fabric of our collectives from top to bottom in order to absorb the citizen of the eighteenth century and the worker of the nineteenth. We shall have to transform ourselves just as thoroughly in order to make room, today, for the nonhumans created by science and technology. (136)

For "[s]o long as humanism is constructed through contrast with the ob-ject that has been abandoned to epistemology," he says, "neither the human nor the nonhuman can be understood" (136). But this posthumanist politics that a posthumanist epistemology can help make possible is not, as Haraway reminds us, a matter of voluntarism; it is not as if having a good attitude and taking thought will restore a hybridized world to the clarity and definiteness of identity for the purposes of political praxis. Indeed, as Haraway points out,

> Most important obligations and passions in the world are unchosen; "choice" has always been a desperately inadequate political metaphor for resisting domination and for inhabiting a livable world. Interpellation is not about choice; it is about insertion. . . . If technological products are cultural actors, and if "we," whoever that problematic invitation to inhabit a common space might include, are technological products at deeper lev-els than we have yet comprehended, then what kind of cultural action will forbid the evolution of OncoMouse™ into Man™? ("When Man™ Is on the Menu" 43)

Pragmatism, Feminist Philosophy of Science, and the Detour of Objectivity

One of the most prominent and important attempts to answer Haraway's question—and to pursue more generally the prospect of posthumanist theory—has been undertaken by feminist philosophy of science, which has sought to ground "cultural action" by attempting to rehabilitate the notion of objectivity. What is paradoxical about this desire for "objectivi-ty" is that it issues from a line of critique that has reminded us again and again that putatively "objective" scientific accounts are just as socially constructed as any other, and that indeed what we might call the ideology of objectivity has typically operated much to the detriment of women and other marginalized people. In a passage justly famous for its candid statement of the contradictory theoretical desires that characterize much feminist philosophy of science at the current moment, Haraway writes,

> I think my problem and "our" problem is how to have *simultaneously* an account of the radical historical contingency for all knowledge claims and knowing subjects, a critical practice for recognizing our own "semiotic technologies" for making meanings, and a no-nonsense commitment to faithful accounts of a "real" world, one that can be partially shared and friendly to earth-wide projects of finite freedom, adequate material abundance, modest meaning in suffering, and limited happiness. ("Situated Knowledges" 187)

There are several important issues at stake in this passage that are crucial for refiguring the relationship between knowledge, ethics, and political praxis for feminism and beyond. But what is most important for my purposes is the linkage between the ethical and political values cataloged at the end of the passage and the "faithful" accounts of a "real world" that should underwrite or otherwise serve as a foundation for the practice of those values. This strategy in Haraway, Fox Keller, Harding and others, I want to argue, is counterproductive because it thrusts the discussion back into a representationalist frame that is both epistemologically inadequate to the task at hand and potentially troubling both politically and ethically.[6]

As I have argued elsewhere,[7] I agree wholeheartedly with Haraway that "the projects of crafting reliable knowledge about the 'natural' world cannot be given over to the genre of paranoid or cynical science fiction," that "social constructionism cannot be allowed to decay into the radiant emanations of cynicism" ("Situated Knowledges" 184) so that—to paraphrase Fox Keller—what counts as knowledge is determined by nothing more than which laboratory has the most money. But I wholeheartedly *disagree* that this means we should redouble our commitment to what Sandra Harding has recently called "strong objectivity"—a leaner and meaner scientific method that would "identify and eliminate distorting social interests and values from the results of research" (17) by "systematically examining all of the social values shaping a particular research process" (18). The problem with Harding's position, of course, is that it assumes that there *is* some space from which to survey our "social interests and values" without at the same time being bound by those interests and values—a space, in other words, of noncontingent observation, a place where one can tally up all of the "blind spots" without having that tally compromised—rendered less than "objective"—by its own blind spot.

Even if Harding wants to break with an "absolute" sense of objectivity that presumes what Richard Rorty calls "a God's-eye standpoint," a "view from nowhere" (6), she does so only to rely upon a "procedural" form of objectivity that assumes that the chaff of "distorting social interests and values" can be objectively separated from the wheat of nondistorting ones.[8] And when one asks, "distorting in relation to *what*?" then it seems (as the ocular figuration of the problem suggests) that we are back within the representationalist frame that fails to acknowledge what the *other* half of Harding, the constructivist half, knows full well: that there is "no way," as Rorty puts it "of formulating an independent test of accuracy of representation—of reference or correspondence to an 'antecendently determinate' reality—no test distinct from the success that is supposedly explained by this accuracy" (*Objectivity* 6). To use the terms of Francisco Varela, Evan Thompson, and Eleanor Rosch in *The Embodied Mind*, Harding's "strong objectivity" is in the end just a form of "weak representationalism"—representationalism with apologies, as it were—because in saying "that different perceiving organisms simply have different perspectives on the world," it "continues to treat the world as pregiven; it simply allows that this pregiven world can be viewed from a variety of vantage points" (*Embodied Mind* 202).

Again, my intention is not to take issue with the admirable *political* values and aims of Harding's project: to argue against the uses of science in promoting inequality and environmental degradation; to critique the reproduction of Eurocentrism and racism in scientific institutions; to open the practice and resources of science as an institution and discipline to marginalized peoples. My point, rather, is to expose the theoretical incoherence of presuming that these values and aims must be grounded in some notion of objectivity. Just as Haraway insists that "only partial perspective promises objective vision" ("Situated Knowledges" 190), so Harding argues that "the systematic activation of democracy-increasing interests and values—especially in representing diverse interests in the sciences when socially contentious issues are the object of concern— in general contributes to the objectivity of science" (18). In response to which one must simply ask how greater diversity of socially "interested" knowledges can add up to a more "objective" sort of knowledge, when objectivity is by definition precisely the sort of knowledge you get once you have removed, rather than expanded, the influence of "social interests and values" upon it? Harding might respond that only "distorting" and "obscuring" interests and values need be removed, but that, as we have seen, is precisely the rub, for who is to say—especially without foreclosing the sort of democratic debate and radical questioning that Harding rightly encourages—when an interest is distorting and when it is not? And if I say that about another, then how is *my* interest not unduly influencing the process?

These difficulties are symptomatic of the essential fallacy at work in the assumption, to borrow Barbara Herrnstein Smith's characterization,

> that objectivism is wrong when practiced by the wrong people for the wrong reasons, but right when practiced by the right people for the right reasons: specifically, that objectivist arguments are culpably "authoritarian" when they issue from powerful agents attempting to justify their own self-interested actions, but laudably "critical" when they issue from disinterested agents exposing the unjust acts of powerful people against subordinated people. Such distinctions, however, are impossible to maintain either theoretically or practically. (295)

While we have already lingered upon the theoretical incoherence diagnosed by Smith, what is not so clear—but every bit as important—are the disabling practical implications mentioned by Smith. For the assumption that there is a necessary correlation between the legitimacy or achievement of the political aims of feminist philosophy of science and the attainment of objectivity ("strong" or not) on the epistemological plane is, I think, rhetorically counterproductive, because it creates a self-defeating contradiction between Harding's polemical, political project (to open up scientific knowledge to "outsider" values and perspectives) and her theoretical, epistemological project (to continue to define what counts as legitimate knowledge by measuring it against a representationalist standard of "objectivity"). To put it another way, Harding's polemical/political project wants to open up science as an institution to social representation, but her theoretical and epistemological premium on "objectivity"—in separating social interests and values from the objects of research, in separating distorting from nondistorting values—only reinforces the disciplinary *insularity* of science as a discursive community from the social space.

"Democratic values," Harding writes, "ones that prioritize seeking out criticisms of dominant belief from the perspective of the lives of the least advantaged groups, tend to increase the objectivity of the results of re-search" (18). But how can such a thing logically be the case? What *is* the case, however, is that such a process, while it has nothing to do with objectivity—except maybe laudably calling the very notion into ever more radical question—*does* have plenty to do with politics. "Representing diverse interests" in the sciences and "seeking out criticisms of dominant belief" in the sciences *do just that*; they don't "achieve the elimination of objectivity-damaging social values and interests" but instead *propagate* those values and interests for the purposes of greater democratic representation of the points of view in the knowledge-making process of Harding and those she presumes to speak for. And that, from a pragmatic point of view, is all that a social and political critique of knowledge *can* do. And in this light (it probably goes without saying) the practical, rhetorical disadvantages of the "strong objectivity" program I have just noted take on renewed significance.

As a range of self-professedly "relativist" theorists have pointed out, it need not be assumed that alternative political and ethical visions must be "grounded" in some objective view of the world, and that only by reference to such objectivity does one have the right to criticize the existing order of things. After all, to tamper with Haraway's formulation, what if it turns out that our objective "faithful accounts of a 'real' world" turn out objectively *not* to be ones "that can be partially shared and friendly to earth-wide projects"? What do we do then—abandon those projects? Certainly not. In fact, as Malcolm Ashmore, Derek Edwards, and Jonathan Potter have recently argued, "Realism is no more secure than relativism in making sure the good guys win, nor even in defining who the good guys are—except according to some specific realist assumptions that place such issues outside of argument" (11). For them, it is the objectivist position that courts political conservatism and quietude, while "it is for relativists and constructionists that the good life is to be lived and *made*, as and in accountable social action, including that of social analysis" (8). Indeed, from Herrnstein Smith's point of view, there may be "a certain grandeur" to objectivist claims, but "[w]hat is sacrificed to obtain that grandeur ... namely, the acknowledgment of both human variability and the mutability of the conditions of human existence—is likely to be paid sooner or later in political coin" (292).

To attempt to ground progressive political praxis in objectivity is then—to borrow Rorty's phrase about Habermas—to "scratch where it does not itch" by attempting to provide a metanarrative of objectivity, rationality, or universalism to ground the contingent "first-order" narratives at work in social life ("Habermas and Lyotard" 164). As Rorty puts it, "The pragmatist's justification of toleration, free inquiry, and the quest for undistorted communication can only take the form of a comparison between societies which exemplify these habits and those which do not, leading up to the suggestion that nobody who has experienced both would prefer the latter. . . . Such justification is not by reference to a criterion"—such as objectivity—"but by reference to various detailed practical advantages" (*Objectivity* 29). This does not mean, as the archrealist or even the representationalist-with-apologies is sure to rush in and exclaim, that the pragmatist is really just an idealist in disguise, a claim whose more radical form is that the pragmatist has no way to show us that the "real" world exists. For this charge, as Ashmore, Edwards, and Potter point out, "trades upon the objectivist assumption that rejecting realism is the same thing as rejecting everything that realists think is real" (8). The pragmatist, Rorty explains, "believes, as strongly as does any realist, that there are objects which are causally independent of human beliefs and desires" (*Objectivity* 101), but in granting this "causal stubbornness" to the world does not grant "an *intentional* stubbornness, an insistence on being *described in a certain way,* its *own* way" (*Objectivity* 83).

Nor does this mean that so-called "facts" of the sort invoked by feminist philosophy of science's realist side are then simply ad hoc constructions driven by, and only by, political expediency. It is true, in pragmatist terms, that nothing prevents us *epistemologically* from going around and making up knowledge claims that seem outlandish; but much prevents us *institutionally* and pragmatically from doing so if we want those claims to receive a serious hearing and count as knowledge within a given discourse. (In a way, this is simply to remind ourselves of the essentially ethical imperative of a certain brand of postmodern neo-Kantianism that insists, in thinkers as otherwise diverse as Lyotard and Habermas, that we respect the separation of discourses and the autonomy of language games.)[9] "Facts," then, as Rorty explains,

> are hybrid entities; that is, the causes of the assertability of sentences include both physical stimuli and our antecedent choice of response to such stimuli. To say that we must have respect for the facts is just to say that we must, if we are to play a certain language game, play by the rules. To say that we must have respect for unmediated causal forces is pointless. It is like saying that the blank must have respect for the impressed die. The blank has no choice, and neither do we. (*Objectivity* 81)

The relativist, constructivist, pragmatic critique, then, doesn't say that the "real world" doesn't exist, that there is no such thing as a "fact," or that we can blithely falsify the data as we go along. It simply means jettisoning the epistemological pretensions that want to ground certain practices and values in "objectivity" and grounding them instead in whether or not they work, as agents of adaptation to an environment, for contingent, revisable purposes. Thus, "From a pragmatist point of view," Rorty writes,

> to say that what is rational for us now to believe may not be *true*, is simply to say that somebody may come up with a better idea. It is to say that there is always room for improved belief, since new evidence, or new hypotheses, or a whole new vocabulary, may come along. For pragmatists the desire for objectivity is not the desire to escape the limitations of one's community, but simply the desire for as much intersubjective agreement as possible, the desire to extend the reference of "us" as far as we can. (23)

On this view, it is perfectly possible to appeal to experimental evidence (as many antirealists do) not because it provides a more "accurate" or transparent reflection of the way things "really are" in the world, but rather because it is persuasive within the rules of knowledge for a given discourse.

The objectivist/realist will no doubt want to challenge this claim by appealing to science's effectivity, but it is quite possible to account for that effectivity by extending the powerful social constructivist arguments mobilized by Latour, Steve Woolgar, and others in studies such as *Science in Action* and *Laboratory Life*. Science, in this view, is privileged not because of its representational transparency to the real, but rather because it *works*. And this fact, in turn—despite the realist attempt to use science's effectivity as evidence of the freedom of science's truth claims from the arena of social power and political rhetoric—only foregrounds the imbrication of science in that very arena, for the question we then must ask is, "works for *whom*, for what purposes?" In this context, it makes sense, of course, that feminist philosophy of science would want to trade upon the considerable rhetorical power of "objectivity" to affect social and institutional change. But the problem, as we have seen, is that their claims for "objectivity" are made not within a rhetorical, political frame—in which one cunningly appropriates notions that are philosophically suspect because they carry powerful appeal for specific audiences (other feminists, say, who are not philosophers or literary theorists)—but are offered instead squarely within an epistemological investigation of the status of knowledge claims. And if one then wants to ask, "so who cares about epistemology?" the answer must be that *we* care—and so do Haraway, Harding, and Fox Keller, who, after all, write epistemological books for theoretical, academic audiences. If we want to meet the epistemological critique of objectivity by devaluing epistemology itself as being "academic" in the worse sense, we must remember Ashmore et al.'s reminder: "But *we* are academics, for whom it is proper, even essential, to care about the epistemic and ontological status of claims to knowledge" (9).

What I am suggesting, then, is that the pragmatic dimension of recent feminist philosophy of science be disengaged from the objectivist epistemological pretensions that undercut its political and ethical commitment and that it epistemologically retool its critique so that it can coherently theorize the paradoxical desire that it voices again and again: for a contingency that is not myopic, a constructivism that is more than just self-serving stories.

It could be argued by Haraway, Harding, and others in response to this suggestion that the second-order cybernetics we are about to examine cannot very well critique the use of "objectivity" and at the same time offer *itself* as a transdisciplinary paradigm that claims universal descriptive validity. But the rejoinder, as we shall see, is that if we agree that *all* critiques or theories are reductive of the verticality of difference (because they are all contingent, which means in turn that another distinction can always be drawn, i.e., we could have distinguished things otherwise), then the issue becomes how to build the confrontation with that fact into the epistemology one is using, rather than continuing to pretend that this contingency and paradoxicality doesn't exist by strategically repressing it.

In the meantime, to avoid constantly undercutting their political critique with an epistemology ill-equipped to serve it, when Haraway in "Situated Knowledges" says "objectivity" she should instead say what she really means, which is "situatedness" and "responsibility," and when Harding says "objectivity" she should instead just say "democracy" and "representation of marginalized voices." This will be difficult for feminist philosophy of science to do, because it is, after all, philosophy of *science*. But once it has affected this disengagement, it will have much to teach pragmatists like Rorty, whose complacent ethnocentrism, as I have argued elsewhere,[10] needs to be confronted with the more muscular pragmatism that is alive and well in Haraway, Harding, and Fox Keller, the latter of whom, in her recent study *Secrets of Life, Secrets of Death*, puts squarely on the front burner the sort of question often avoided or blithely glossed over by pragmatism in its Rortyan incarnation. "From critical theory, to hermeneutics, to pragmatism," she writes,

> the standard response to so-called relativist arguments has been that the scientific stories are different from other stories for the simple reason that they "work." If there is a single overriding point I want to make . . . it is to identify a chronic ellipsis in these responses: As routinely as the effectiveness of science is invoked, equally routine is the failure to go on to say what it is that science works *at,* to note that "working" is a necessary but not sufficient constraint. (74)

Only by forcing examination of these *specific, material* effects of scientific discourse and practice can we forge a more socially and politically responsive pragmatist critique of knowledge that understands that if science is "what works," it always works *at* something *for a* "particular 'we' . . . embedded in particular cultural, economic, and political frames" (Fox Keller 5). Only by paying this sort of attention can we force the pragmatist commitment to contingency to take itself at its word and undertake a full critique of what Fox Keller calls the "romance of disembodiment" on not only epistemological grounds, but political ones as well.

When Loops Turn Strange: From First- to Second-Order Cybernetics

In light of the posthumanist imperative I have been invoking thus far, systems theory has much to offer as a general epistemological system. Unlike feminist philosophy of science, it does not cling to debilitating representationalist notions. And unlike Enlightenment humanism in general, its formal descriptions of complex, recursive, autopoietic systems are not grounded in the dichotomy of human and nonhuman. Indeed, in the posthumanist context I have sketched above, the signal virtue of systems theory is, as Dietrich Schwanitz puts it, that it has "progressively undermined the royal prerogative of the human subject to assume the exclusive and privileged title of self-referentiality (in the sense of recursive knowledge about knowledge)" (267). Hence, systems theory promises a much more powerful and coherent way to describe the complex, intermeshed networks of relations between systems and their specific environments of whatever type, be they human, animal, ecological, technological, or (as is increasingly the case) all of these.

But there exists within systems theory itself an important distinction between first- and second-order cybernetics that we will need to understand before we can grasp the full originality and importance of the second-order cybernetics of Humberto Maturana and Francisco Varela. To get a sense of *first*-order cybernetics and why posthumanist theory must move beyond it, there is no more instructive example for my purposes than the cultural anthropologist and intellectual polymath Gregory Bateson, who from the 1940s through the 1970s engaged in an ambitious and fascinating attempt to extend the new theoretical model of cybernetics into the social sciences to describe the basic formal dynamics of alcoholism, communication among wolves and dolphins, primitive art and ritual, ecological crisis, schizophrenia, and much else besides.

The fundamental principle of cybernetics in its original formulation during and immediately after World War II is circular causality or "recursivity," a principle whose most well-known name probably remains the "feedback loop," of which there are two types: negative feedback, in which information is processed by the system in such a way as to maintain the harmony or homeostasis of the system, and positive feedback, in which information is processed in a such a way as to destabilize the system and create what is sometimes called a "vicious cycle." Although positive feedback is important to recent work in complexity theory,[11] we must leave it aside to concentrate on negative feedback, a famous example of which is offered by Steve J. Heims in his recent social history of cybernetics:

> A person reaches for a glass of water to pick it up, and as she extends her arm and hand is continuously informed (negative feedback)—by visual or propioceptive sensations—how close the hand is to the glass and then guides the action accordingly, so as to achieve the goal of smoothly grabbing the glass. (15)

What is immediately intriguing about this example of negative feedback—and about the principle of circular causality in general—is that it contains a paradox, one that second-order cybernetics will pursue to its logical conclusions, as first-order cybernetics never really did: A causes B and B causes A. As Heims explains, "The process is circular because the position of the arm and hand achieved at one moment is part of the input information for the action of the next moment" (15–16). And hence, the system is characterized by "recursivity" which, as Niklas Luhmann defines it, is a process that "uses the results of its own operations as the basis for further operations—that is, what is undertaken is determined in part by what has occurred in earlier operations. In the language of systems theory . . . one often says that such a process uses its own outputs as inputs" (72).

Despite its interdisciplinary range and explanatory power, Bateson's work stops short of pursuing the full implications of this paradoxical fact about recursivity (A causes B *and* B causes A), and the contingency of all observation to which such paradoxicality attests (we can say *either* A causes B *or* B causes A, thus it is always possible to observe otherwise). The move from first- to second-order cybernetics is characterized, as Heinz von Foerster argues in *Observing Systems,* by the full disclosure of this epistemological problem:

> 1. observations are not absolute but relative to the observer's point of view (i.e., his coordinate system—Einstein's theory or relativity); 2. observations affect the observed so as to obliterate the observer's hope of prediction (i.e., his uncertainty is absolute—Heisenberg's uncertainty principle). Given these changes in scientific thinking, we are now in the possession of the truism that a description (of the universe) implies one who describes (observes it). (Quoted in Kenney and Boxer 76)

What is most intriguing about Bateson's work is that on the one hand he wants to insist in essays like "Redundancy and Coding" and "Cybernetic Explanation" on the contingency of observation, on the constructivist point that the sort of knowledge you get depends upon the code or map that you use—that "the map is not the territory," to the phrase of Korzybski's that he often quotes (*Steps* 449). But on the other hand, that recognition of contingency gets undone by Bateson's totalizing insistence that there is a single loop or overarching *"pattern which connects"* observer and observed (*Mind and Nature* 8), so that what looks at first glance like contingent observation is instead determined "from behind" by the total pattern of existence, generating what Bateson calls an immanent "mental determinism" (*Steps* 465). This is quite clear in later essays like "Form, Substance, and Difference," where Bateson writes,

> The cybernetic epistemology which I have offered you would suggest a new approach. The individual mind is immanent but not only in the body. It is immanent also in pathways and messages outside the body; and there is a larger Mind of which the individual mind is only a subsystem. This larger Mind is comparable to God and is perhaps what some people mean by God, but it is still immanent in the total interconnected social system and planetary ecology. (*Steps* 460)

It is at moments like these that the epistemological rigor of second-order cybernetics proves decisive and invaluable. As von Foerster suggests, the crucial realization of second-order cybernetics is that you cannot have your constructivist contingency—regardless of whatever liberating ethical or political implications might flow from it—and eat it too. For once it is acknowledged that observation is contingent (i.e., could be otherwise), then it must also be acknowledged that total loops such as those imagined by Bateson must always turn into "strange" loops of the sort imaged by M. C. Escher's Möbius strip. For as Ranulph Glanville and Francisco Varela remind us in their elegant little demolition of total loops entitled "Your Inside Is Out and Your Outside Is In (Beatles [1968])," the distinction between inside and outside, system and environment, mind and nature, always contains a paradox that makes the distinction turn back upon itself to form a strange loop. This is so, they argue, because when we draw any putatively final distinction in either intension or extension—when we attempt to distinguish either the elementary or universal—"we require that its distinction has no inside and, at the same time we place, in this non-existent inside a further distinction which asserts that the distinction of the fundamental was the last distinction!" (639). Thus, they continue, "at the extremes we find there are no extremes. The edges dissolve BECAUSE the forms are themselves continuous—they re-enter and loop around themselves" (640), not like a Batesonian circle of the total system, but like a Möbius strip, a more fitting image for the paradoxicality of distinction—a paradoxicality that, second-order cybernetics forces us to say, *must always accompany the assertion of the contingency of the observer, of the fact that an observation could always be otherwise.*

This abandonment of the total "pattern which connects" on behalf of the contingency of observation and the sort of systemic heterogeneity it makes recognizable links second-order cybernetics rather directly to broader currents of postmodern theory such as that practiced by Deleuze and Guattari. "I part company with Bateson," Guattari writes,

> at the point where he defines action and enunciation as mere segments of the ecological sub-system known as context. . . . There is no overall hierarchy of enunciative ensembles and their sub-sets, whose components can be located and localized at particular levels. Those ensembles are made up of heterogeneous elements which acquire consistency and persistence only as they cross the thresholds that bound and define one world against another. They are . . . [like] Schlegel's "little works of art." ("Like a little work of art, a fragment has to be totally detached from the surrounding world and closed upon itself like a hedgehog.") (141)

Or, as we are about to see, like the autopoietic organizations of second-order cybernetics which—far from participating in an "immanent determinism" driven by the total "pattern which connects"—are totally self-referential because they exist by virtue of what Maturana and Varela will call their "operational closure." Under the sign of second-order cybernetics and its postmodern cognates, knowledge appears instead, in Varela's words,

> more and more as built from small domains, that is microworlds and microidentities. . . . [S]uch microworlds are not coherent or integrated into some enormous totality regulating the veracity of the smaller parts. It is more like an unruly conversational interaction: the very presence of this unruliness allows a cognitive moment to come into being according to the system's constitution and history. ("Re-Enchantment" 336)

Here, then, we find the full implications of second-order cybernetics' emphasis on the contingency of observation, its constant reminder, as Maturana and Varela put it, that "everything that is said is said by someone." Because all contingent observations are made by means of the "strange loop" of paradoxical distinction between inside and outside, x and not-x, "every world brought forth," they write, "necessarily hides its origins. By existing, we generate cognitive 'blind spots' that can be cleared only through generating new blind spots in another domain. We do not see what we do not see, and what we do not see does not exist." (*Tree* 242)

Between the Scylla of Realism and the Charybdis of Idealism: Autopoiesis and Beyond

The key distinction for the theory of autopoiesis (or "self-production") as articulated by Maturana and Varela—the distinction that, as we shall see in a moment, allows its decisive conceptual innovation, its account of systems that are both open and closed—is the distinction between "organization" and "structure." As they explain it, "*Organization* denotes those relations that must exist among the components of a system for it to be a member of a specific class"; it is that which "signifies those relations that must be present in order for something to exist." *Structure,* on the other hand, "denotes the components and relations that actually constitute a particular unity and make its organization real" (*Tree* 46, 47). For example, the basic and necessary *organization* of the water-level regulation system in a toilet consists of a float and a bypass valve. But in terms of the *structure,* for example, the float that is made of plastic could be replaced by one made of wood "without changing the fact that there would still be," as Maturana and Varela somewhat infelicitously put it, "a toilet organization" (*Tree* 46). This basic distinction between organization and structure will mark a crucial epistemological innovation in their attempt, as they put it, to "walk on the razor's edge, eschewing the extremes of representationalism (objectivism) and solipsism (idealism)" (*Tree* 241). It will also, more broadly, enable a reconceptualization of the relationship between *system* (organization + structure) and *environment* (everything outside the system's boundaries), which will mark a definitive break with the first-order cybernetics of Bateson.

For Maturana and Varela, what characterizes all living things is that they are "*autopoietic organization[s],*" that is, "they are continually self-producing" (*Tree* 43) according to their own internal rules and requirements. In more general terms, what this means is that all autopoietic entities are *closed*—or, to employ Niklas Luhmann's preferred term, "self-referential"—on the level of *organization,* but open to environmental perturbations on the level of *structure.* This is clearest, perhaps, in Maturana and Varela's contention that all autopoietic entities are defined by "*operational closure.*" "It is interesting to note," they write,

> that the operational closure of the nervous system tells us that it does not operate according to either of the two extremes: it is neither representational nor solipsistic.
>
> It is not solipsistic, because as part of the nervous system's organism, it participates in the interactions of the nervous system with its environment. These interactions continuously trigger in it the structural changes that modulate its dynamics of states....
>
> Nor is it representational, for in each interaction it is the nervous system's structural state that specifies what perturbations are possible and what changes trigger them. (*Tree* 169)[12]

Environmental "triggers" and "perturbations," then, take place on the level of structure, but what may be *recognized* as a perturbation or trigger is specified by the entity's organization and operational closure. What this means, Maturana and Varela conclude squarely against the first-order cybernetics of Bateson, is that the model of the nervous system "picking up information" from the environment is misleading (*Tree* 169). "[I]nformation," as Varela, Thompson, and Rosch put it in *The Embodied Mind,* is not "a prespecified quantity, one that exists independently in the world and can act as the input to a cognitive system." After all, they ask, "how are we to specify inputs and outputs for highly cooperative, self-organizing systems such as brains?" (139). The difference between cognitive systems—and, Maturana and Varela would argue, autopoietic systems in general—and input/output devices is, in the words of Marvin Minksy, "that brains use processes that change themselves—and this means we cannot separate such processes from the products they produce" (*Embodied Mind* 139).

Here, then, we can see how second-order cybernetics radicalizes the concept of recursivity abandoned prematurely by first-order cybernetics. As we have seen, first-order cybernetics avoids the crude representationalism and realism that holds, as Richard Rorty puts it, that "'making true' and 'representing' are reciprocal relations: the nonlinguistic item which makes S true is the one represented by S" (*Objectivity* 4). But it does so only to smuggle representationalism back in in the form of the input-output model and the notion of "information-processing." For Maturana and Varela, revealing the poverty of the representational frame for making sense of such phenomena as perception, color vision, cognition, and memory is absolutely crucial to their entire epistemological project, which aims to "negotiate a middle path between the Scylla of cognition as the recovery of a pregiven outer world (realism) and the Charybdis of cognition as the projection of a pregiven inner world (idealism)." "These two extremes," Varela et al. contend, "both take representation as their central notion: in the first case representation is used to recover what is outer; in the second case it us used to project what is inner" (*Embodied Mind* 172).

And at this juncture, Maturana and Varela in *The Tree of Knowledge* broach the question that any antirepresentationalist epistemology sooner or later must confront: namely, the question of relativism. "If we deny the objectivity of a knowable world," they ask, "are we not in the chaos of total arbitrariness because everything is possible?" The way "to cut this apparent Gordian knot," they respond, is to realize that the first principle of any sort of knowledge whatsoever is that "everything said is said by someone"—to foreground, in short, the problem of observation (*Tree* 135). As Varela et al. put it, "Our intention is to bypass entirely this logical geography of inner versus outer by studying cognition not as recovery or projection but as embodied action" (172)—"embodied" because cognition depends upon the "individual sensorimotor capacities" of the embodier in *situ,* and "active" (or "enactive") because the cognitive structures that guide perception and action—as dramatically demonstrated by the example of color vision—"emerge from the recurrent sensorimotor patterns that enable action to be perceptually guided" (173).[13] The full definition of "embodiment," then, is a self-referential, self-organizing, and nonrepresentational system whose modes of emergence are made possible by the history of structural coupling between the autopoietic entity and an environment to which it remains closed on the level of organization but open on the level of structure. This cluster of terms constitutes what Varela calls a "radical paradigmatic or epistemic shift," which holds that the lived, concrete, contingent, embodied quality of all knowledge "is not 'noise' that occludes the brighter pattern to be captured in its true essence, an abstraction, nor is it a step toward something else: it is how we arrive and where we stay" ("Re-Enchantment" 320).

But this acknowledgment of the full complexity of autopoietic systems does not dispense with systematic description altogether. Instead, it recasts the relationship between a system and its elements (or, to use the language of Maturana and Varela, an organization and its structure) as open-ended and yet not random, fundamental and yet not foundational in the usual ontological sense. As Dietrich Schwanitz puts it,

> the elements function as units only within the system that constitutes them, they are neither just analytical constructs nor do they rest in some ontological substance. They really do exist, but their existence is only brought about by self-reference and cannot in any way be explained by reference to preexisting ideas, substances or individuals. (272)

This loss of meaning (if one wants to put it in that representational way) is, according to Varela, totally unavoidable, and nowhere is this clearer than in his work on perception and cognition, which reveals the temporal structure of the cognitive transition from one moment or action to the next to be extremely "fine" in texture, consisting of a "fast dynamics" or "fast resonance" of neuronal activity in which we find extremely rapid cooperation and competition between distinct neural agents ready to constitute different frames of action and interpretation of the perceptual event. "On the basis of this fast dynamics," Varela explains,

> as in an evolutionary process, one neuronal ensemble (one cognitive sub-network) finally *becomes more prevalent and becomes the behavioral mode for the next cognitive moment.* By "becomes more prevalent" I do not mean to say that this is a process of optimization: it resembles more a bifurcation or symmetry-breaking form of chaotic dynamics. *It follows that such a cradle of autonomous action is forever lost to lived experience* since, by definition, we can only inhabit a microidentity when it is present, not when it is in gestation. ("Re-Enchantment" 334, second emphasis mine)

The particular suppleness of this sort of descriptive apparatus, then, is that it provides us with "a philosophical system, a reductive system," as Varela et al. put it, "in which reductive basic elements are postulated as

ultimate realities but in which those ultimate realities are not given ontological status in the usual sense" (*Embodied Mind* 118).

More than a few readers have suggested that this way of negotiating
the realism/idealism problem constitutes a kind of double-dealing—a
cooking of the books of nature, you might say. The Marxist sociologist
Danilo Zolo, for example, has suggested recently that a persistent confusion about the claims and status of autopoiesis haunts the work of
Maturana and Varela. On the one hand, Zolo argues, they want to maintain a last, fretful tie to empiricism. They go out of their way to claim that
the theory of autopoiesis does not rely upon reference to forces or dynamics "not found in the physical universe" (as they put it in *Autopoiesis
and Cognition*), that autopoietic unity "is not an abstract notion of purely conceptual validity for descriptive purpose, but is an operative notion"
(quoted in Zolo 67, 68). But on the other hand, they want to espouse a
thoroughgoing constructivist position that holds that any scientific explanation is always, as they put it, "a reformulation of a phenomenon,"
that when we describe an autopoietic system "we project this system
upon the space of our manipulations and make a description of this projection" (quoted in Zolo 67). As Zolo sees it, Maturana and Varela want
to hold that predictions about what happens in physical space (as opposed to the abstract and conceptual domain) are valid because, as they
put it, "a description, as an actual behavior, exists in a matrix of interactions which (by constitution) has a logical matrix necessarily isomor-
phic with the substratum matrix within which it takes place" (quoted in
Zolo 69). But this, Zolo argues, only redoubles the contradictory status of
the claims of autopoiesis. "They forget," Zolo writes,

> that they have already argued that it is impossible to distinguish "between
> perception and hallucination in the operation of the nervous system" . . .
> that nothing can be said about the "substratum" of observation; that
> knowledge has no object and that everything that can be said is always said
> by an observer. Thus, it is meaningless to postulate the existence of a "logi-
> cal isomorphism" between the substratum of the observation and the lan-
> guage of description. (69)

The problem foregrounded but not fully understood, I think, by Zolo's
critique—nor, it should be added, always clearly articulated by Maturana
and Varela—is one we have already mentioned: the problem of *observation*. In a recent essay, Maturana offers what is in effect a response to
Zolo's critique—and in particular to Zolo's undertheorized mobilization
of the dichotomies objective/subjective, realist/idealist, and so on:

The fact that science as a cognitive domain is constituted and validated in the operational coherences of the praxis of living of the standard observers as they operate in their experiential domains without reference to an independent reality, *does not make scientific statements subjective.* The dichotomy of objective-subjective pertains to a cognitive domain in which the objective is an explanatory proposition that asserts, directly or indirectly, the operational possibility of pointing to an independent reality. Science does not, and cannot, do that. ("Science and Daily Life" 41–42, emphasis mine)

But this response only foregrounds the need to theorize even more rigorously the concept of observation. "As observers," Maturana and Varela write, "we can see a unity in *different* domains, depending on the distinctions we make"; we can consider the internal states and structures of a system, or we can consider how that system interacts with its environment. For the former observation, "the environment does not exist"; for the latter, "the internal dynamics of that [system's] unity are irrelevant" (*Tree* 135). The key point, then, is that

> both are necessary to complete our understanding of a unity. It is the observer who correlates them from his outside perspective. It is he who recognizes that the environment can trigger structural changes in it. It is he who recognizes that the environment does not specify or direct the structural changes of a system. The problem begins when we unknowingly go from one realm to the other and demand that the correspondences we establish between them (because we see these two realms simultaneously) be in fact a part of the operation of the unity. (*Tree* 135–36)

Maturana in a recent essay offers an even more nuanced explanation of his concept of observation, one that helps us to see how Zolo's critique is mounted upon a foundation of epistemological reductionism. In Maturana's view, by contrast, the

> nonreductionist relation between the phenomenon to be explained and the mechanism that generates it is operationally the case because the actual result of a process, and the operations in the process that give rise to it in a generative relation, intrinsically take place in independent and nonintersecting phenomenal domains. This situation is the reverse of reductionism; scientific explanations as generative propositions constitute or bring forth a generative relation between otherwise independent and nonintersecting phenomenal domains, which they thus de facto validate. ("Science and Daily Life" 34)

What this means, I take it, is that the scientific explanation or observation constitutes the *relation* between "the phenomenon to be explained" (the observer's view of the system in its environment, which is not possible from the vantage of the system) and the "mechanism" or "operations" (the relation between the system's operationally closed organization and its structure, which is open to environmental triggers).

The key words here, then, are "actual" and "nonintersecting"; the "result of a process" is "actual," not only because it is what the observer *sees*, but also because—as we have already seen in our discussion of emergence— the descriptive specification she chooses to make in her observation is *binding* with regard to how the "generative" processes—the relation between system and environment, system and element, organization and structure—can be construed. Once the observer has specified the system in question in her account of the phenomenon, the generative relations between organization and structure in the system being observed are *not* random or whimsical but must in fact be systematic. All of which is to say that the observation and explanation of a phenomenon constitutes, de facto validates, and in this sense "generates" the relationship between the observed phenomenon (the "actual result of a process" of system plus environment) and operations of the system that give rise to it. But the crucial point here is that the phenomenon and those generative operations take place in "nonintersecting domains" that become joined—but also potentially confused—in the scientific explanation. As Maturana and Varela put it above, "The problem begins when we unknowingly go from one realm to the other"—from the vantage of the environment to that of the system, both of which are joined in the observed "phenomenon to be explained"—"and demand that the correspondences we establish between them (because we see these two realms simultaneously) be in fact a part of the operation of the unity" (*Tree* 135–36).

The Persistence of Humanism: Maturana and Varela's Ethics

The second-order cybernetics of Maturana and Varela, as it turns out, has much the same lesson to learn from feminist philosophy of science as Rortyan pragmatism. This does not mean, however, that it is subject to the most common critiques leveled at the Janus-faced politics of first-order cybernetics by feminist philosophers of science such as Haraway, ecological feminists such as Carolyn Merchant, and popular social critics such as Jeremy Rifkin.[14] Merchant's critique is standard: that the systems theory paradigm can "be appropriated, not as a source of cultural transformation, but as an instrument for technocratic management of society and nature, leaving the prevailing social and economic order unchanged" (104). Indeed, as Steve J. Heims points out in his recent social history of the Macy cybernetics conferences of 1946–53, the conferences themselves were conducted in the stringently apolitical atmosphere of the Cold War, which hung over first-order cybernetics as a whole, one in which questions of politics, ideological differences, and alternative social configurations were strongly discouraged, if not forbidden.

But if these sorts of critiques may be valid for first-order cybernetics, it is difficult to see how they would hold for second-order cybernetics, with its emphasis upon the radical contingency of observation, the embodiment of knowledge, and the irreducible complexity of systemic description that flows from both. As we have already seen, second-order cybernetics, by pursuing the full implications of the principle of recursivity held at bay in its predecessor, concerns itself at least as much with the creative and unpredictable capacities of self-organizing and autopoietic systems as with the mechanisms of control and closure foregrounded by the Macy Conferences' emphasis on systemic homeostasis. And while second-order systems theory does make a claim to universal descriptive veracity, that claim is mounted upon its ability to theorize the *inability* to see the social or natural system as a totality from any particular observer's point of view. It is difficult, therefore, to see how second-order cybernetics could justly be described as in principle a theoretical instrument of globalized "technocratic management" when it foregrounds the very contingency, complexity, and unpredictability that such programs of technocratic control would want to repress, ignore, or deny.

It is more useful—and more apropos the theoretical commitments of second-order cybernetics—to reframe the work of Maturana and Varela in terms of what Merchant calls the need for "reconstructive knowledge," which should be based on "principles of interaction (not dominance), change and process (rather than unchanging universal principles), complexity (rather than simple assumptions), contextuality (rather than context-free laws and theories), and the interconnectedness of humanity with the rest of nature" (107). If it seems far-fetched to read the second-order cybernetics of Maturana and Varela in this light, we should remember that they themselves have cast the pragmatic and ethical import of their theoretical work very much in these terms. As they put it at the end of *The Tree of Knowledge,*

> The *knowledge of knowledge compels.* It compels us to adopt an attitude of permanent vigilance against the temptation of certainty. . . . It compels us

to realize that the world everyone sees is not *the* world but *a* world which we bring forth with others. It compels us to see that the world will be different only if we live differently. (245)

Maturana and Varela understand, as does feminist philosophy of science in its own way, that the stakes over the epistemological status of "objectivity" are far from purely epistemological. But Maturana and Varela base the ethical and pragmatic value of their work squarely upon the difference between the epistemology of representationalism and realism ("knowledge") retained by feminist philosophy of science, and one that starts from the second-order theorization of the problematics of contingent observation, the fact that "everything said is said by someone" ("knowledge of knowledge"): "*We affirm that at the core of all the troubles we face today is our very ignorance of knowing.* It is not knowledge, but the knowledge of knowledge, that compels" (*Tree* 248). The "knowledge of knowledge" leads Maturana and Varela to now conclude, in a quite remarkable passage, that second-order cybernetics "implies an ethics we cannot evade":

> If we know that our world is necessarily the world we bring forth with others, every time we are in conflict with another human being *with whom we want to remain in coexistence,* we cannot affirm what for us is certain (an absolute truth) because that would negate the other person. If we want to coexist with the other person, we must see that *his certainty—however undesirable it may seem to us—is as legitimate and valid as our own.* . . . Let us not deceive ourselves; we are not moralizing, we are not preaching love. We are only revealing the fact that, biologically, without love, without acceptance of others, there is no social phenomenon. (*Tree* 246–47)

It is hard to imagine a more powerful statement of the ethical imperatives of second-order cybernetics than this.

Unfortunately, it is also hard to imagine a more powerful symptom of the unreconstructed humanism that is just as inadequate (if not more so) to the epistemological innovations of second-order cybernetics as the epistemology of feminist philosophy of science is to its progressive political agenda. That humanism here manifests itself in the philosophical idealism that hopes that ethics may somehow do the work of politics. What we find here, in other words, is (to borrow Fredric Jameson's formulation) a kind of "strategy of containment" whereby the posthumanist imperatives of second-order cybernetics are ideologically recontained by an idealist faith in the social and political power of reason, reflection, voluntarism, and what Jameson calls "the taking of thought" (*Political Unconscious* 52–53, 59–60, 282–83): *"We affirm that at the core of all the troubles we face today is our very ignorance of knowing."*

My point is not to take issue with their emphasis on the importance of "bringing forth a common world" (indeed, as we shall see, the distinctly utopian cast of their formulation of the problem links them, interestingly enough, to the utopianism of Jameson's Marxism, which takes as a kind of ultimate given the "hankering after collectivity" at work in the social project). My point is simply to remind us, as Jameson puts it in *The Political Unconscious,* that ethical thought "projects as permanent features of human 'experience,' and thus as a kind of 'wisdom' about personal life and interpersonal relations, what are in reality the historical and institutional specifics of a determinate type of group solidarity or class cohesion" (59). It is precisely this contradiction that lies buried in Maturana and Varela's crucial but subordinated proviso, *"with whom we want to remain in coexistence,"* which solves before the fact, as it were, the problem of social (and for him economic and class) difference that Jameson highlights. Maturana and Varela's ethical assertion of the necessity of love (aside from being a paradoxical imperative that commands "you *will* love!" to a subject who "freely" chooses) is predicated upon the assumption that the question of "with whom we want to remain in coexistence" has always already been solved. In the process, Maturana and Varela drain the assertion of contingency of its materialist, pragmatic force, whose entire point—as we know from feminist philosophy of science as well as Marxist theory—is to say that all points of view are *not* equally valid precisely because they have material *effects* whose benefits and drawbacks are distributed asymmetrically in the social field. And this asymmetry, in turn, makes it vastly easier for some groups and persons to enjoy the luxury of freely accepting the "validity" of points of view other than their own. This, after all, is the point of Fox Keller's assertion that the practice of knowledge always works *at* something specific and for a particular *we.* Maturana and Varela are right that, epistemologically speaking, all points of view are equally contingent; but this does not mean, from a pragmatic point of view, that we need treat all points of view as as equally "legitimate and valid." Indeed, as Ashmore et al. point out, "if objective truth and validity are renounced in favor of social process and practical reasoning, then so also must be any notion of a commitment to '*equal* validity.' Far from ruling out the possibility of justification of a particular view, relativism insists upon it" (10).

Such advice seems even more crucial to remember in light of the use to which Buddhist philosophy is put in *The Embodied Mind*. In chapter 10 of that study, for example, Varela et al. want to distinguish their Buddhist commitment from Western pragmatism proper, and they argue that "Western philosophy has been more concerned with the rational under- standing of life and mind than with the relevance of a pragmatic method for transforming human experience" (218). But what becomes clear in later chapters is that this "pragmatic method" consists of repeated calls for us to heed the wisdom of Buddhist "mindfulness" and "egolessness" to solve by ethical fiat and spiritual bootstrapping the complex problems of social life conducted in conditions of material scarcity, economic inequality, and institutionalized discrimination of various forms. This is especially clear in their critique of Garret Hardin's "The Tragedy of the Commons," where they respond to the problem of scarcity and the self- interested conduct it generates in terms already familiar from *The Tree of Knowledge*: "We believe that the view of the self as an economic man, which is the view the social sciences hold, is quite consonant with the un- examined view of our own motivation as ordinary, non-mindful people" (246). And the "pragmatic" answer to self-interested conduct created by conditions of economic scarcity, they tell us, is *not* to address that materi- al scarcity and inequality itself, but rather to encourage through enlight- enment "an attitude of all encompassing, decentered, responsive, com- passionate concern," which "must be developed and embodied through a discipline that facilitates letting go of ego-centered habits and enables compassion to become spontaneous and self-sustaining" (252).

But clearly, as we have already suggested, this amounts to little more than telling people that the problems of scarcity and the maldistribution of wealth and power will go away if we all simply stop being so selfish— a claim which is very easy for some to make and very hard for some to hear. Here, as elsewhere in Maturana and Varela, the complicated rela- tionship between ethics and politics is not so much explained as ex- plained away by an appeal to total human transformation with little or no attention to the material factors that make that appeal little more than wishful thinking. And from this vantage, "love" as Maturana and Varela define it can in fact be *anti*social, *even if* it preserves "the biologic process that generates" the social process. In the end, then, Maturana, Varela et al. give us "embodiment," but not a robust, *socially and historically situated* embodiment, and their "pragmatism" is disabled by exactly what they criticize in Husserl: that the "self" and its "experience"—the linchpins of their critique of formalist epistemology—remain "entirely *theoretical*" and lack any "*pragmatic* dimension" (*Embodied Mind* 19). As Vincent Kenny and Philip Boxer put it in their comparison of Maturana and Lacan, "What *does* make the difference between the family, the asylum and the concentration camp as forms of social structural coupling? If there are those who would argue that these are all the fruits of reflection and an 'opening up of room for existence,' are reflection and love enough therefore as an ethics?" (95).

The answer would seem to be "no," not only for Jamesonian reasons but also, as it were, for post-Jamesonian ones: that Maturana and Varela's call for an ethic of love constitutes a radical disavowal of what Ernesto Laclau, Chantal Mouffe, and Slavoj Zizek have called "social antagonism."[15] Maturana and Varela, in their call for acceptance of the other's view as valid no matter what it is, engage in what Zizek calls a fetishistic disavowal of antagonism whose form is "I know very well there are views which I despise, but still . . ." ("Beyond Discourse-Analysis" 259). "What this fetishistic logic of the ideal is masking," Zizek writes, "is, of course, the limitation proper to the symbolic field as such: the fact that the signifying field is always structured around a fundamental deadlock" that is "not covered by any ideal (of the unbroken communication, of the invention of the self)" (259). What Maturana and Varela disavow is nothing other than the "auto-negativity" and "self-hindering" status of the subject and its desire, its lack, its traumatic "internal limit." Indeed, "the stake of the entire process of subjectivation, of assuming different subject-positions," Zizek writes, "is ultimately to enable us to avoid this traumatic experience" (253). As Zizek puts it,

> "the subject" in the Lacanian sense is the name for this internal limit, this internal impossibility of the Other, of the "substance." The subject is a paradoxical entity which is, so to speak, its own negative, i.e., which persists only insofar as its full realization is blocked—the fully realized subject would be no longer subject but substance. (254)

We will remember that the Lacanian name for this substance is, of course, the Real or what Kant called in the *Critique of Practical Reason* the "pathological" Thing, *das ding*. And in this light, it becomes clear that Maturana and Varela's terrifying injunction, "Love!," is in reality a call for the end of desire, the need to repress the Thing at the heart of the subject, the "biology," if you will, at the heart of the "biological process."

And when we recall, moreover, that the most familiar name for substance, *das ding*, and the Real since Freud's *Civilization and Its Discontents* is the *animal*, it becomes painfully clear that the surest sign of Maturana and Varela's persistent humanism is not their individualism, nor even their idealism, but rather the systematic *speciesism* that is unmistakable in their work separately and in collaboration. It is not simply that Maturana and Varela frame their ethics solely in terms of the reciprocal relations between *human beings,* and in doing so undercut the promise of their epistemology by leaving aside the very posthumanist imperatives—of ecology, of animal rights, of the political and ethical challenges of techno-science—which we mentioned at the beginning. It is rather the jarring, symptomatic contradiction upon which their ethical project runs aground again and again: on the one hand it rigorously demonstrates that the human species is not the only one to participate in social, cultural, and linguistic domains (if not languaging proper), and recognizes the unique-ness and importance for social organization and communication among nonhuman animals of individual temperament and ontogeny—all of which are factors that, for them, constitute grounds for ethical considera-tion if we apply consistently the rules that they use for human ethical con-sideration.[16] But on the other hand, their work systematically invokes and praises some of the most invasive and brutal animal research on monkeys, cats, rabbits, and other animals conducted in recent decades.

This quintessentially humanist "blind spot" constitutes an almost un-bearable myopia in Varela et al.'s *The Embodied Mind,* where the authors call for "the cultivation of compassion for all sentient beings" (248), for a "responsiveness to oneself and others as sentient beings without ego-selves" (251). Having issued such a call, they then proceed to praise the extremely controversial neurophysiological research of Russell DeValois on macaque monkeys (170 ff.) (which has been challenged for nearly a decade for its brutality and frivolity by several leading animal rights groups) and recount a "beautiful study" in which kittens were raised in the dark, kept entirely passive, and as a result when released "after a few weeks of this treatment," acted "as if they were blind: they bumped into objects and fell over edges" (175).

This blindness on the part of the authors, however, will perhaps come as less of a surprise when we remember that the relationship between subject and substance in the Enlightenment paradigm as articulated by Zizek is one of traumatic disavowal of the bond between meaning and substance, self and thing, human and animal. In this light, the surest sign of humanism is that

> subjectivation designs the movement through which the subject integrates what is given him/her into the universe of meaning—[but] this integra-tion always ultimately fails, there is a certain left-over which cannot be

integrated into the symbolic universe, an object which resists subjectivation, and the subject is precisely the correlative to this object. ("Beyond
Discourse-Analysis" 254)

Maturana and Varela hope that "love" will achieve such an integration,
but it is clear that the most quintessentially humanist "left-over" in their
discourse, as in humanism generally, is the animal other as articulated by
the discourse of speciesism, with the subject of humanism its precise correlative. Maturana and Varela's humanist ethics thus fails precisely because it *is* humanist; it attempts to solve by ethical fiat the posthumanist
political challenges that their epistemology might help us to theorize.
Their ethics forgets what their epistemology knows: that in the cyborg
cultural context of OncoMouse™ and hybrids of nature/culture, the
question is not who will get to be human, but what kinds of couplings
across the humanist divide are possible and indeed unavoidable when we
begin to observe the end of Man.

Notes

1. See, for example, Lentricchia's chapter on Foucault in *Ariel and the Police*.

2. As Zizek notes in *The Sublime Object of Ideology,* "Habermas and Foucault
are two sides of the same coin"; "the Foucauldian notion of subject enters the
humanist-elitist tradition" by way of the later Foucault's notion of the subject as
"mastering the passion within himself and making out of his own life a work of
art," whereby we find a realization of "subject as the power of self-mediation and
harmonizing the antagonistic forces, as a way of mastering the 'uses of pleasure'
through a restoration of the image of self" (2).

3. I have argued this point elsewhere. See my "Nature as Critical Concept." In
this connection, see also Benton and Soper.

4. See Jameson, *Late Marxism,* especially 20–21, 35–36, 96–99, 214–15; and
Cornell, especially 16–24.

5. In cognitive ethology, see Griffin and the essays collected by Bekoff and
Jamieson. In field ecology, see Goodall, and Cheney and Seyfarth. In cognitive
science and philosophy of mind, see Dennett and Dawkins. And in animal rights
philosophy, see Regan and Singer.

6. The desire to hold on to the concept of objectivity is not by any means
limited to feminist philosophy of science. See, for example, Levine; and see
Lenoir's discussion of a similar project in the work of Bruno Latour.

7. See Wolfe, "Making Contingency Safe for Liberalism."

8. I borrow these characterizations of different types of objectivity from
Megill's editorial introduction.

9. As is well known, that discursive difference and autonomy is subjected to a rather different fate in the end by Habermas and Lyotard, with the former insisting upon the adjudication of knowledge claims by different discourses by the process of rational consensus, and the latter insisting that the intractable "differends" between those different language games be respected, even at the price of abandoning any hope for consensus. For an overview, see Best and Kellner.

10. See Wolfe, "Making Contingency Safe for Liberalism."

11. See, for example, the work of economist Brian Arthur, who has worked extensively with the Santa Fe Institute on complexity theory. For a useful popular account, see Waldrop.

12. This view, in fact, is widely held in neurobiology (the scholarly field of research of Maturana and Varela) and in cognitive science, where philosophers such as Daniel Dennett agree with Maturana and Varela that "our world of colored objects is literally independent of the wavelength composition of the light coming from any scene we look at. . . . Rather, we must concentrate on understanding that the experience of a color corresponds to a specific pattern of states of activity in the nervous system which its structure determines" (*Tree* 21–22). See Dennett, especially the chapter "Qualia Disqualified" (which contains a section entitled "Why Are There Colors?"). For further discussion of the example of vision by Maturana and Varela, see *Tree* 18–23, 126–27, and 161–62.

13. For a more detailed account, see Varela, "The Re-Enchantment of the Concrete" 332–35.

14. See Haraway's "Sex, Mind, and Profit"; Merchant's *Radical Ecology*; and Rifkin's *Algeny*.

15. This theory of antagonism would take issue not only with Maturana and Varela, but with Jameson's positing—in *The Political Unconscious* and essays like "Reification and Utopia in Mass Culture"—of a utopian hankering after collectivity and the projection of an "external enemy"—call him the capitalist—"who is preventing me from achieving identity with myself," when in reality this projection of an "other which is preventing me from achieving my full identity with myself is just an externalization of my own auto-negativity, of my self-hindering" which can never be abolished, "come the revolution" or otherwise (Zizek, "Beyond Discourse-Analysis" 252–53).

16. See in particular chapters 8 and 9 of *The Tree of Knowledge*, especially 212, 224.

Works Cited

Adorno, Theodor. *Negative Dialectics*. Translated by E. B. Ashton. New York: Seabury, 1973.

Ashmore, Malcolm, Derek Edwards, and Jonathan Potter. "The Bottom Line: The Rhetoric of Reality Demonstrations." *Configurations* 1 (1994): 1–14.

Bateson, Gregory. *Mind and Nature: A Necessary Unity.* New York: Bantam, 1988.

———. *Steps to an Ecology of Mind.* New York: Ballantine, 1972.

Bekoff, Marc, and Dale Jamieson, eds. *Interpretation and Explanation in the Study of Animal Behavior.* Vol. 1. Boulder: Westview, 1990.

Benton, Ted. "The Malthusian Challenge: Ecology, Natural Limits, and Human Emancipation." In *Socialism and the Limits of Liberalism,* edited by Peter Osborne. London: Verso, 1991.

Best, Steven, and Douglas Kellner. *Postmodern Theory: Critical Interrogations.* New York: Guilford, 1991.

Cheney, D. L., and R. M. Seyfarth. *How Monkeys See the World.* Chicago: University of Chicago Press, 1990.

Cornell, Drucilla. *The Philosophy of the Limit.* New York: Routledge, 1992.

Dawkins, Marian Stamp. *Through Our Eyes Only?: The Search for Animal Consciousness.* Oxford: Freeman, 1993.

Dennett, Daniel. *Consciousness Explained.* Boston: Little, Brown, 1991.

Foucault, Michel. "Nietzsche, Genealogy, History." In *The Foucault Reader,* edited by Paul Rabinow. New York: Pantheon, 1984.

———. "Truth and Power." In *The Foucault Reader,* edited by Paul Rabinow. New York: Pantheon, 1984.

Glanville, Ranulph, and Franciso Varela. "Your Inside Is Out and Your Outside Is In (Beatles [1968])." In *Applied Systems and Cybernetics.* Proceedings of the International Congress on Applied Systems Research and Cybernetics. Vol. 2, *Systems Concepts, Models, and Methodology.* Edited by G. E. Lasker. New York: Pergamon, 1980.

Goodall, Jane. *Through a Window: My Thirty Years with the Chimpanzees of Gombe.* Boston: Houghton Mifflin, 1990.

Griffin, Donald. *Animal Minds.* Chicago: University of Chicago Press, 1992.

Guattari, Félix. "The Three Ecologies." *New Formations* 8 (summer 1989): 131–47.

Haraway, Donna J. "A Cyborg Manifesto: Science, Technology, and Socialist-Feminism in the Late Twentieth Century." In *Simians, Cyborgs, and Women: The Reinvention of Nature.* New York: Routledge, 1991.

———. "The Biological Enterprise: Sex, Mind, and Profit from Human Engineering to Sociobiology." In *Simians, Cyborgs, and Women: The Reinvention of Nature.* New York: Routledge, 1991.

———. "Situated Knowledges." In *Simians, Cyborgs, and Women: The Reinvention of Nature.* New York: Routledge, 1991.

―――. "When Man™ Is on the Menu." In *Incorporations*, edited by Jonathan Crary and Sanford Kwinter. New York: Zone, 1992.

Harding, Sandra. "Eurocentric Scientific Illiteracy—A Challenge for the World Community." Introduction to *The "Racial" Economy of Science*, edited by Sandra Harding. Bloomington: Indiana University Press, 1993.

Heims, Steve Joshua. *Constructing a Social Science for Postwar America: The Cybernetics Group 1946–1953.* Cambridge: MIT Press, 1993.

Jameson, Fredric. *Late Marxism: Adorno, or, The Persistence of the Dialectic.* London: Verso, 1990.

―――. "Reification and Utopia in Mass Culture." *Social Text* 1 (winter 1979): 130–48.

―――. *The Political Unconscious: Narrative as a Socially Symbolic Act.* Ithaca: Cornell University Press, 1983.

Keller, Evelyn Fox. *Secrets of Life, Secrets of Death: Essays on Language, Gender and Science.* New York: Routledge, 1992.

Kenny, Vincent, and Philip Boxer. "Lacan and Maturana: Constructivist Origins for a Third-Order Cybernetics." *Communication and Cognition* 25, no. 1 (1992): 73–100.

Latour, Bruno. *We Have Never Been Modern.* Cambridge: Harvard University Press, 1993.

Lenoir, Timothy. "Was the Last Turn the Right Turn? The Semiotic Turn and A. J. Greimas." *Configurations* 1 (1994): 119–36.

Lentricchia, Frank. *Ariel and the Police: William James, Michel Foucault, Wallace Stevens.* Minneapolis: University of Minnesota Press, 1988.

Levine, George. "Why Science Isn't Literature: The Importance of Differences." In *Rethinking Objectivity*, edited by Allan Megill. Durham: Duke University Press, 1994.

Luhmann, Niklas. "The Cognitive Program of Constructivism and a Reality That Remains Unknown." In *Selforganization: Portrait of a Scientific Revolution*, edited by Wolfgang Krohn, Gunter Kuppers, and Helga Nowotny. Dordrecht: Kluwer, 1990.

Maturana, Humberto R. "Science and Daily Life: The Ontology of Scientific Explanations." In *Research and Reflexivity*, edited by Frederick Steier. London: Sage, 1991.

Maturana, Humberto, and Francisco J. Varela. *The Tree of Knowledge: The Biological Roots of Human Understanding.* Rev. ed. Translated by Robert Paolucci. Boston: Shambhala, 1992.

Megill, Allan. "Four Senses of Objectivity." Introduction to *Rethinking Objectivity*, edited by Allan Megill. Durham: Duke University Press, 1994.

Merchant, Carolyn. *Radical Ecology.* New York: Routledge, 1993.

Osborne, Peter, ed. *Socialism and the Limits of Liberalism*. London: Verso, 1991.

Regan, Tom. *The Case for Animal Rights*. Berkeley: University of California Press, 1983.

Rifkin, Jeremy. *Algeny*. New York: Penguin, 1984.

Rorty, Richard. "Habermas and Lyotard on Postmodernity." In *Habermas and Modernity*, edited by Richard Bernstein. Cambridge: MIT Press, 1985.

———. *Objectivity, Relativism, and Truth*. Vol. 1 of *Philosophical Papers*. Cambridge: Cambridge University Press, 1991.

Schwanitz, Dietrich. "Systems Theory and the Environment of Theory." In *The Current in Criticism: Essays on the Present and Future of Literary Theory*, edited by Clayton Koelb and Virgil Lokke. West Lafayette, Ind.: Purdue University Press, 1987.

Singer, Peter. *Animal Liberation*. New York: Avon, 1975.

Smith, Barbara Herrnstein. "The Unquiet Judge: Activism without Objectivism in Law and Politics." In *Rethinking Objectivity*, edited by Allan Megill. Durham: Duke University Press, 1994.

Soper, Kate. "Greening Prometheus: Marxism and Ecology." In *Socialism and the Limits of Liberalism*, edited by Peter Osborne. London: Verso, 1991.

Spivak, Gayatri. "Remembering the Limits: Difference, Identity, and Practice." In *Socialism and the Limits of Liberalism*, edited by Peter Osborne. London: Verso, 1991.

Varela, Francisco. "The Re-Enchantment of the Concrete." In *Incorporations*, edited by Jonathan Crary and Sanford Kwinter. New York: Zone, 1992.

Varela, Francisco, Evan Thompson, and Eleanor Rosch. *The Embodied Mind: Cognitive Science and Human Understanding*. Cambridge: MIT Press, 1993.

Waldrop, M. Mitchell. *Complexity: The Emerging Science at the Edge of Order and Chaos*. New York: Simon and Schuster, 1992.

Wolfe, Cary. "Making Contingency Safe for Liberalism: The Pragmatics of Epistemology in Rorty and Luhmann." *New German Critique* 61 (winter 1994): 101–27.

———. "Nature as Critical Concept: Kenneth Burke, the Frankfurt School, and 'Metabiology.'" *Cultural Critique* 18 (spring 1991): 65–96.

Zizek, Slavoj. "Beyond Discourse-Analysis." Afterword to *New Reflections on the Revolution of Our Time*, by Ernest Laclau. London: Verso, 1990.

———. *The Sublime Object of Ideology*. London: Verso, 1989.

Zolo, Danilo. "Autopoiesis: Critique of a Postmodern Paradigm." *Telos* 86 (winter 1990–1): 61–80.

III

SYSTEMS THEORY IN RESONANCE WITH

MAJOR POSTMODERNISTS

9. The Limit of Modernity:
Luhmann and Lyotard on Exclusion

William Rasch

Jean-François Lyotard's *The Postmodern Condition* is a deeply divided work. On the one hand, it ceremoniously rejects the so-called Enlightenment projects of modernity, the metanarratives, as the famous phrase has it, of emancipation and knowledge. Thus, the condition labeled "postmodern" paradoxically recognizes the fact that no great alternative, no absolute knowledge or historical subject is waiting in the eschatological wings to transform modernity into its utopic other. On the other hand, however, this recognition does not transform Lyotard into a champion of the modernization process or an apologist for the "system." Like Horkheimer and Adorno, whose analyses of the inescapable horrors of immanence echo throughout *The Postmodern Condition*, Lyotard retains a profound distaste for what remains after the great alternative projects have failed. Accordingly, modernity as the "administered society" (Horkheimer and Adorno) or as the "performativity of the system" (Lyotard) can only be seen as a hell on earth that is compounded by the absence of any messianic promise of salvation. The solution to this dilemma—i.e., the simultaneous need for, and lack of, an outside—is to posit the hope for an immanent (albeit not imminent) self-transformation of modernity. If Horkheimer and Adorno endeavored to move beyond the concept by way of the concept, Lyotard attempts to move beyond the performativity of science by way of its paralogy. The other of the present, therefore, is said to arrive not from the outside, as revelation or apocalypse, but parasitically from within, as an ethical imperative that is housed in what the system excludes or marginalizes. This ethical redemption of the system is said, then, to compensate for the collapse of the political projects of the utopic alternative. What remains problematic about this solution, however, is not the notion that a body—a "system"—can carry a "subversive" parasite, but that this parasite can be considered a moral agent or can otherwise be ethically steered.

Lyotard's relationship to Niklas Luhmann, as fleeting and oblique as it has been, registers his ambivalence both about the "unsurpassable horizon" of the ever-expanding "interior space of modernity" (Wellmer vii) and about the possibility of an ethic of the excluded other. If, in *The Postmodern Condition,* Luhmann is linked, via Parsons, to Comte, and made to stand for totality, efficiency, and terror (12, 61–64), by the late 1980s he has come to be seen more as an ally than an enemy. In response to a meeting with Luhmann in 1988, Lyotard writes:

> N.L., hardly loquacious, calculating his words with his Baudelairian elegance (which is much more than a systematic strategy), knows this kind of complexity. He wants to simplify a different kind of complexity. And he can only do it at the cost of a supplement of differentiation. It is this apparent "aporia," assumed with calm and tact, that I like most in his thought, from which I am in a sense so distant. It was possible for us to form a small common front against the waves of ecologist eloquence. A two-sided front. There is no nature, no *Umwelt,* external to the system, he explained. And I added: of course, but there remains an *oikos,* the secret sharer *[hôte]* to which each singularity is hostage. (*Political Writings* 81)[1]

Lyotard now stresses a mutual rejection of an accessible outside. The affirmation of immanence is no longer recorded by Lyotard as an indicator of efficiency, but rather appreciated for its paradox, the "aporia" of a necessary "supplement of differentiation," or, as Luhmann would say, the reentry of the system/environment distinction within the system. However, that the system itself can produce effects that can only make their presence felt as disturbances, as if they had come from without, is not only for Lyotard a logical imperative, but also, as his emotive language suggests ("secret sharer," "hostage"), a last residue or possibility of ethical action. The system produces noise; Lyotard wants to hear that noise as a "call." It is on this point that a profitable *Auseinandersetzung* between the two can proceed. Consequently, this paper pursues the question of whether that which presents itself as the other of a system—and ultimately that which presents itself as the other of modernity—is to be thought of as its logical limit or its moral conscience.

In a most succinct, if indirect, manner, Lyotard expresses the dilemma felt by many who no longer feel tempted by the call for a radical transformation of society. "All politics," he writes, "is only (I say 'only' because I have a revolutionary past and hence a certain nostalgia) a program of administrative decision making, of managing the system" (*Political Writings* 101). With a touch of self-deprecatory irony, he acknowledges the collapse of a two-hundred-year, Enlightenment/Marxist tradition of oppositional politics, yet distances himself from what remains in the aftermath. This dual gesture is key to understanding his attempt to think both the inevitability and the possible instability of modernity. Already in *The Postmodern Condition,* the famous demise of the metanarratives of emancipation and *Bildung* signals the rejection of this political tradition. Even the opposition between "traditional" and "critical" theory no longer holds, Lyotard realizes, because the Archimedian efforts of critical theory have disintegrated into postulates of "utopia" and "hope." In essence, critical theory, which has variously grounded itself in some historical subject (the proletariat, and then, "the Third World or the students"—and one can easily add to Lyotard's list) or in categories like "man or reason or creativity," has lost its claim to be able to occupy an outside or oppositional position and has therefore become just one more regulator of the system. In "countries with liberal or advanced liberal management," it has, in so many words, become co-opted, while in communist countries it has become the system itself. Thus: "Everywhere, the Critique of political economy (the subtitle of Marx's *Capital*) and its correlate, the critique of alienated society, are used in one way or another as aids in programming the system" (*Postmodern Condition* 13).

Dismantling the eschatology of emancipation remains a theme throughout the writings of the 1980s and early 1990s. In "Rewriting Modernity," Lyotard attacks the hermeneutics of remembering, "as though the point were to identify crimes, sins, calamities engendered by the modern set-up—and in the end to reveal the destiny that an oracle at the beginning of modernity would have prepared and fulfilled in our history" (*Inhuman* 27). In "The Wall, the Gulf, and the Sun: A Fable," oppositional criticism and the interest in emancipation, far from opposing the system from the outside, are seen as necessary means by which the system improves its efficiency (*Political Writings* 113–14). But perhaps most telling, and poignant, is his 1989 introduction to a republication of his essays on the Algerian war for independence, essays that were originally written during his association with the group "Socialism or Barbarism" in the 1950s and 1960s. Here, the demise of the Enlightenment/Marxist political project is delineated with great clarity:

The presumption of the moderns, of Christianity, Enlightenment, Marxism, has always been that another voice is stifled in the discourse of "reality" and that it is a question of putting a true hero (the creature of God, the reasonable citizen, or the enfranchised proletarian) back in his position as subject, wrongfully usurped by the imposter. What we called "depoliticization" twenty-five years ago was in fact the announcement of the erasure of the great figure of the alternative, and at the same time, that of the great founding legitimacies. This is more or less what I have tried to designate, clumsily, by the term "postmodern." (*Political Writings* 169)

Yet, with this acknowledgment of the total collapse of the project of emancipation, Lyotard is faced with a dilemma: If one rejects the traditional/critical opposition as outdated, and if one rejects the historical narratives of emancipation and knowledge from which this opposition could gain nourishment, where does one turn if one wishes to escape the deadening embrace of what Lyotard variously calls the system, the monad, or the ethos of development? If one can no longer think the disenfranchised other as the site for oppositional political activity, is, then, the attempt to think the other bereft of all significance? Lyotard is certainly not claiming that the problem of exclusion in the form of political oppression has disappeared, or that exclusion is now somehow to be preferred. Rather, he simply observes that the inclusion of the excluded (the proletariat, the Third World, women, and so forth) as the subject of history can no longer be proposed as the basis of an emancipatory political program. The challenge becomes, then, one of thinking exclusion in ways not compromised by utopian projections of the great alternative.

The dynamic logic of exclusion is an inherent feature of Luhmann's systems theory, a feature that has become increasingly highlighted with reference to George Spencer Brown. Spencer Brown's "laws of form" serve as a refinement, a logical shorthand, for the enforced selectivity that is the hallmark of Luhmann's notion of complexity and thus of his notion of system formation. All choice, all observation—as the act of making distinctions, of making "cuts" in the world—is a process of inclusion by way of exclusion. As Luhmann explicitly points out in a recent essay:

The concept of form designates the postulate that operations, insofar as they are observations, always designate and actualize one side of a distinction and mark it as the starting point for further operations—not the other side which, as it were, is carried along empty *(leer mitgeführt wird)*. . . . The theoretical provocation of this concept of form . . . rests on its postulate that by virtue of an operation's coming into being something is always excluded—at first in a purely factual manner, but then as something logically necessary for an observer capable of making distinctions. ("Inklusion" 240)

[Der Begriff der Form bezeichnet dann das Postulat, daß Operationen, soweit sie Beobachtungen sind, immer die eine Seite einer Unterscheidung bezeichnen, aktualisieren, als Ausgangspunkt für weitere Operationen markieren—und nicht die andere Seite, die im Moment gleichsam leer mitgeführt wird. . . . Die theoriebildende Provokation des Formbegriffs . . . beruht darauf, daß er postuliert, daß durch das Zustandekommen einer Operation immer etwas ausgeschlossen wird—zunächst rein faktisch, sodann aber für einen Beobachter, der über die Fähigkeit des Unterscheidens verfügt, logisch notwendig.]

Interesting here is the distinction between factual and logical necessity. On the level of operations, exclusions are by-products of an enforced selection, a reduction of complexity, an identity formation. A system—living, social, or other—defines itself against a background, which, as its environment, remains inaccessible. We start off, as it were, in a room with two doors. When we walk through one (marked, for instance, "male"), the other door ("female") disappears from view. The door not chosen "wird leer mitgeführt," meaning that we can never walk back out again, only through additional doors, which may now be marked differently, but we, by now, are marked differently too. One can walk through the "male" door (or, more precisely, be walked through it at birth), and then, if one happens to be of a romantic habit of mind, attempt to think androgyny, but it will always be a male-centered androgyny, an androgyny "seen" from the perspective of one who initially entered the "male" door, and who therefore "carries with him" the rejected "female" door as a permanent blind spot.[2]

As an operation, all this remains rather unproblematic. It is the way of the world. Controversy—that is, choice of perspective—arises on the level of observation, indeed, the level from which the above description was made. Exclusion was presented as a logical necessity, not just a factual occurrence. Of necessity, choice precludes other possibilities. By way of the inclusion/exclusion distinction, observation sees that the operation of observation includes what it chooses, and excludes what it does not. Seen from this "logical" point of view, exclusion is presented as unavoidable. Just because one can observe the excluded as excluded does not mean that the excluded can now be painlessly included, for this logical observation also operates by way of exclusion and can only see a former exclusion, a "latency," by way of a new exclusion. Try as we might, we have not developed alternative logics, ones that could promise exclusion-free inclusion. Thus, *remediating* the effects of the process of exclusion can only happen by *replicating* the effects of the process of exclusion.

Such a depiction of logical inescapability, however, raises more than a few hackles and takes on a different shading when a moral or political distinction is substituted for, or superimposed over, the logical one. From a political perspective, the excluded becomes the other (of the system, of the dominant discourse, and so forth). If we stick with the example used above, we can see that in a patriarchal society, the male "self" awards himself the attributes of an assumed universality (i.e., desired traits like strength, rationality, educability, seriousness), while the female "other" becomes the source of unwanted (or, more rarely, idealized) deviance (their obverse: weakness, irrationality, "natural" immutability, frivolous-ness). More than a logical necessity, exclusion is thus read as a series of existential consequences of ideological choice. From such a political per-spective, to maintain a logical or scientific *(wissenschaftlich)* observation of the logical necessity of exclusion is deemed an evasion or denial of the victimized other, if not, in fact, a further masculinist strategy of domina-tion. Indeed, according to this view, logic itself, by hiding (excluding) the political analysis, becomes ideological. If a whole culture, in the name of humanism, is walked through the door called "Man," and if "Man," not so coincidentally, bears a striking resemblance to "man," then the logical ex-clusion is no longer merely the way things are, but rather, the way things have deliberately been made to be and, therefore, the way things ought not to be. Thus, to remain "neutral" about the excluded other is tanta-mount to a moral affirmation of that which is included, the privileged self; and this affirmation, it is felt, must be met with critique in the name of the excluded.

With impeccable severity, Luhmann opposes the political reading of exclusion, referring to it as a "victimology." Within this "victimological" tradition, the excluded *(das Ausgeschlossene)* is personified as a class, or as some other form of human collectivity, and mourned. "Were society to respond as demanded to this complaint," he maintains,

> it would still not become a society that excluded nothing. It would com-municate out of other considerations, with other distinctions, and per-haps resolve the paradoxes of its communication differently, shift sor-row and pain and, by doing so, create a different silence. ("Speaking and Silence" 36)

The pathos of personification is simply no match for the inexorable grinding of the logic of exclusion. Inclusion, even the inclusion of the oppressed other, is predicated on exclusion. Such an observation need not be construed as irrevocably hostile to particular political activity. Oppressed minorities and exclusionary ideologies undeniably exist, as well as laws and legal systems that are inherently and systemically prejudicial with regard to the rights of select groups. And political activity in an attempt to rectify perceived injustices and inadequacies is a part of our daily lives. But all this does not erase the logical fact, Luhmann argues, that a politics that would claim to give voice to the excluded other for the sake of egalitarian inclusivity is a constitutive impossibility.

Therefore, if we attempt to think both the logic and the politics of exclusion, we find ourselves in the presence of what Lyotard calls a differend. "A case of differend between two parties takes place," he explains, "when the 'regulation' of the conflict that opposes them is done in the idiom of one of the parties while the wrong suffered by the other is not signified in that idiom" (*Differend* 9). In our case, however, even identifying the differend seems to be inextricably entwined with the differend we try to describe. If we say that the wrong suffered by women is not signified in the neutrally logical idiom of inclusion/exclusion, we would, of course, already assume the position of the political. And if we say that the differend registers the indeterminate conflict between incommensurable language games, between the logical or descriptive versus the political or prescriptive, then the very attempt to describe the nature of the conflict would be a descriptive gesture, hence, a gesture that participates in the dispute it seeks to describe. Thus, we find ourselves here at an impasse. More precisely, we find ourselves replicating the Kantian antinomy between theoretical (descriptive) and practical (prescriptive) reason; and in the modern world of unresolved antinomies, there is no *Aufhebung*.

Perhaps because of his "revolutionary past," Lyotard feels the need to deal with the necessity of the differend by way of a kind of "anamnesis," a mournful nonforgetting of the mechanism by which forgetting happens, because to forget that forgetting happens is to fall victim to the beautiful illusion of reconciliation, to a type of Hegelian sublation that claims nothing is left behind, all is remembered in a transformed, higher stage of knowledge. This rejection of *Aufhebung* brings Lyotard back to the Kantian starting point of the antinomy between theoretical and practical reason, that is, the impossibility of deducing a prescription from a description, which leads to the impossibility of cognitively justifying an ethical "call" or obligation received from an unknown source (*Differend*

107–27). For Lyotard, as was the case with Kant, an ethical observation is autonomous, not derived from, and therefore not subordinated to, knowledge. Autonomy, however, does not mean isolation, it means perpetual conflict, a battle in which each side—theory and practice—attempts to assert the hegemony of its own observer-position. This continuous struggle, however, must effectively remain a stalemate if we are to be true to the differend. We can represent the situation as follows: We start with an event (or, if we wish to speak with Luhmann, an operation) by generating both a theoretical and a practical observation of the event, and these two observations stand in an incommensurable relationship to each other. We mark this relationship as a first-order differend—differend$_1$—and then proceed to observe this first-order differend from both a theoretical and practical observer-position. These observations of observations likewise stand in an incommensurable relationship to each other and, therefore, we can mark this relationship with the term differend$_2$, and say that this second-order differend replicates the structure of its first-order cousin and in no way resolves the dispute.[3] If we have lost faith in logical resolutions by way of neutral third terms, then we see that syntheses of these observations are, in fact, translations of differends into litigations, i.e., successful adjudications of disputes by phrasing one idiom in terms of the other. A synthesis would in fact be an ethical domination of the theoretical, or a logical domination of the practical; it would not resolve the original incommensurability, but only render it invisible. Syntheses are decisions that mask themselves as the *avoidance* of decisions. They thus compound the "violence" (the exclusion) that all decisions make under the pretense of excluding violence.

Does this mean that the structure of the differend leads to paralysis? No, for choice is necessary. If the above sketch of the import of the differend has any meaning, then Lyotard's injunction to "bear witness to the differend" (*Differend* xiii) can be read as an attempt to make the ineluctable violence of enforced selectivity visible.[4] The question to be asked, then, is the following: Is the imperative to bear witness to the differend a practical observation of the differend, or a theoretical one? Is, in other words, the imperative to acknowledge the necessity and necessary violence of choice an ethical or a logical imperative?

We can find evidence for both options in Lyotard. Certainly there is a strong moral, even religious flavor to much of his writings, especially with regard to the Holocaust. Indeed, Lyotard attempts to use the figure of "the jews" as an abstract marker not only for real Jews, who have, quite literally and in a variety of quite violent ways, been excluded from European civilization over the centuries, but also for

> all those who, wherever they are, seek to remember and to bear witness to something that is constitutively *forgotten*, not only in each individual mind, but in the very thought of the West. And it [the expression "the jews"] refers to all those who assume this anamnesis and this witnessing as an obligation, a responsibility, or a debt, not only toward thought, but toward justice. (*Political Writings* 141)[5]

In a more Freudian register, "the jews" stand for the unrepresentable originary shock of the West, its constitutive exclusion, and the "Final Solution was the project of exterminating the (involuntary) witnesses to this forgotten event and of having done with the unpresentable affect once and for all, having done with the anguish that it is their task to represent" (*Political Writings* 143). Given this evocation, the demand to "witness" re-reads the logical necessity of exclusion as The Fall. Anamnesis serves, then, as a ritualized nonforgetting of primordial forgetting, of Original Sin. It is not that lost innocence could be thereby regained—that would be the eschatological project of emancipation—but that it would be remembered and mourned with every choice, every exclusion, every reenactment of the original fall from grace. But can this act of memory escape the forgetting it mourns? Does it not already also replicate the exclusion that would be the object of the justice it demands? One could ask whether the figure "the jews" does not attempt the type of reconciled neutrality Lyotard knows is not possible. Do not "the jews" evoke real Jews, and thus exclude other potential markers for the other of the West—"the native americans," say, or "the africans," or "the homosexuals?"[6] To raise these questions, of course, implies observations just as morally and politically charged as the observations they question. That, in fact, would be the logical point of asking them in the first place.[7]

There is, however, a way of reading Lyotard that is stripped, or nearly so, of the theological pathos. It relies on a different distinction than the one between the theoretical and the practical, and it brings Lyotard back in closer association with Luhmann. Lyotard is of course famous for his resurrection of a particular kind of Kantian aesthetic, especially for his championing of the notion of the sublime. The distinction of importance in this regard is the one between determinate judgment—constitutive of conceptual knowledge—and reflective judgment—constitutive of the nonconceptual aesthetic response to particularity. Even though the sublime can be linked, in both Kant and Lyotard, with ethical thinking, I would like to elucidate the determinate/reflective distinction in terms of function and nonfunction. The attempt here is to think the limit of function (or *Zweckmäßigkeit*) and thus to think the limit of modernity itself.

"The differend cannot be resolved," Lyotard writes. "But it can be felt as such, as differend. This is the sublime feeling" (*Lessons* 234). One hesitates to conceptualize the sublime, because here the sublime—a "feeling"—marks the limit of conceptualization. It does not describe the differend, because as we have seen, any such description already phrases the dispute in one of the contentious idioms, or masks itself with some supposedly neutral third term. Instead, the sublime marks the limit, the incommunicable other of communication, the event that announces nothing, causes nothing. The sublime represents no lost or transcendent *Ding an sich*, but rather presents itself as the realization of the either/or of limits, the distinction that irrevocably cuts the world, making it visible in the very same moment it makes it invisible. Simply and most directly put, the sublime presents the impossibility of thinking the sublation of antinomies. Thus, it would not be overly dramatic to say that Lyotard's notion of the sublime registers the pain of the "severed and mutilated condition" of Spencer Brown's universe, a universe that cuts itself in two to observe itself (105). "Reflection thus touches on the absolute of its conditions," Lyotard observes,

> which is none other than the impossibility for it to pursue them "further": the absolute of presentation, the absolute of speculation, the absolute of morality.... The consequence for thought is a kind of spasm. And [Kant's] Analytic of the Sublime is a hint of this spasm.... It exposes the "state" of critical thought when it reaches its extreme limit—a spasmodic state. (*Lessons* 56)

Lyotard's appropriation of Kant's Third Critique, then, is certainly not made for the sake of a bridge to be built between theoretical and practical reason, nor simply for a way to return to the ethical as something like the gravitational pull of the Law (though this is always present in Lyotard too). Rather, the sublime provides an emphatic demonstration of the radical impossibility of such a bridge and the constitutive impossibility of any straightforward reclamation of victims for the sake of politics or morality. With the sublime, we do not have an observation of the excluded, but a "feeling" of, and for, the mechanism of exclusion. What this "feeling" consists of might best be put as the attempt not to think either side of the differend, neither to conceptualize nor to mourn it. Thus, the import of the sublime resides in drawing the line, not in overstepping it.

We cannot communicate this "feeling," for then it would cease to be a "feeling." We can, nevertheless, choose to conceptualize it and thereby retreat from the limit, confronting the dilemma of the differend once again. If the sublime marks the cut that both enables observation and guarantees that observation will always be partial, casting shadows along with its light, then we find ourselves returning to the logic of exclusion with which we began. Our reflections on the necessity of exclusion have all taken place within the immanent space of a modernity that is marked by untranscendable differends, i.e., exclusions. By definition, there can be no point of indifference from which this space can be thought of as an undifferentiated unity. However, by the very same logic of exclusion that has got us here, must not this immanent space also be a limited space? Must not there be an other to modernity that is contained, as an exclusion, within modernity itself? How are we to think this other of modernity as modernity's limit?

Lyotard offers the following fable. "In the incommensurable vastness of the cosmos," he writes, there are "closed, isolated systems" called galaxies and stars. These systems are marked by entropy, the gradual decrease in "internal differentiation." However, within the vastness of the universe, there are also pockets of negentropy, pockets of unexpected and improbable increases of complexity. "With the advent of the cell, the evidence was given that systems with some differentiation were capable of producing systems with increased differentiation according to a process that was the complete opposite of that of entropy." In time, that recursive and self-referential mode of communication called "language" developed, allowing for "improbable forms of human aggregation . . . according to their ability to discover, capture, and save sources of energy." Finally, the inner differentiation of these social aggregations into "social, economic, political, cognitive, and representational (cultural) fields" led to "systems called liberal democracies," whose task was to control "events in whatever field they might occur. By leaving the programs of control open to debate and by providing free access to the decision-making roles, they maximized the amount of human energy available to the system" (*Political Writings* 120–23).

This more recent fable "updates" the one offered in *The Postmodern Condition* and gives evidence of a more sympathetic reading of Luhmann. The "performativity of the system," once decried as "terrorist," turns out to have been the only viable model available, with no "parology" left to oppose it. Yet the system still has its other. The distinction entropy/negentropy is reentered into the negentropic space, to the effect that within the sea of entropy (the cosmos) we find an island of negentropy (the earth), on which pockets of entropy can be found. Perhaps these pockets can be labeled the "Third World" (*Political Writings* 99), or perhaps they can be found in the form of the unconscious (*Political Writings* 100). But no matter what form it takes, entropy

> is an otherness that is not an *Umwelt* at all, but this otherness in the core of the apparatus. We have to imagine an apparatus inhabited by a sort of guest, not a ghost, but an ignored guest who produces some trouble, and people look to the outside in order to find out the external cause of the trouble." (*Political Writings* 100)

So entropy comes to stand for the limit of development, an internally generated "spasm," with no other function than to be the other of function itself.

Surprisingly, Luhmann has also recently found a hidden guest lodged within the heart of modernity. More surprisingly still, he locates this guest in the Third World and in the ghettos or *favelas* of the large cities in industrialized nations like Brazil and the United States. Unlike the new social movements—which, according to Luhmann, have the specific function of pointing out the failings of functional differentiation[8]— these pockets of exclusion are neither utopic nor dystopic alternatives to functional differentiation, but areas of "negative integration," so to speak, "because the exclusion from one function system quasi-automatically effects the exclusion from others" ("Inklusion" 259).[9] What emerges from this exponential process of exclusion is a form of "supercoding," a superimposition of the inclusion/exclusion distinction over modernity that contradicts the logic of functional differentiation ("Inklusion" 260). Exclusion, of course, is no less a feature of functional differentiation than it is of anything else, but since differentiation manages exclusion by way of system reference and not globally, the type of negatively integrated, total exclusion represented by the *favelas* contradicts modernity's own self-understanding, making the "improbability" and "artificiality" of functional differentiation visible ("Inklusion" 260). Thus, even for Luhmann, the existential reality of exclusion—i.e., the excluded observed as *Personen*—is registered and correlated with the logical necessity of distinction. Like all designations of a distinction, functional differentiation must have its opposite term. Luhmann has traditionally displaced the other of functional differentiation in history as segmentation and/or stratification. But now, from within modernity itself, modernity's other emerges as a logical necessity and a limit function of function itself. What modernity cannot "modernize" returns as a violent spasm, indicating that even functional differentiation, the great ethos of development, has its other and its limit. Even *it* excludes as it includes, and even evolution can make no sense of this process. Modernity's other, it turns out, is neither "pre-" nor "post-."

One is reminded, here, of Carl Schmitt's indictment of liberal universalism. "As long as a state exists," he writes, "there will thus always be in the world more than just one state. A world state which embraces the entire globe and all of humanity cannot exist" (53). Perhaps the same can be said of the "state" *(status)* of modernity as well. It too needs its "adversary." But now that the Western eschatalogical imagination has run its course—for the time being, at any rate—attempts to invest modernity's self-generated other with political or moral authority (as subject of history, as epistemologically privileged "slave" or margin) appear quixotic at best. If we do without these political readings of the limit of modernity, then we reconfigure that limit simply as a logical space.[10] As such, it becomes the latency that houses the excluded as potentiality. Perhaps, when actualized, these potentialities can be felt as disturbances coming as if from the outside, but since we have walked through the door marked modernity, and since we carry that marking with us, the disturbances that actualized potentialities may cause can only be disturbances that are felt within modernity, within the supplement of differentiation that both creates the space beyond its limit and is created by it. Modernity cannot be transformed into its other by such spasms; it can only be extended. Given the spread of fundamentalist revivals of antimodern sentiment around the world, it is fittingly ironic to realize that this logical observation of modernity and its limit cannot help but be a political one as well.

Notes

1. See also the differing ways Luhmann and Habermas are compared in *The Postmodern Condition* (66) and the essay "Oikos" in *Political Writings* (101).

2. Speaking in the language of *Social Systems,* one can say that choice, as structure, is reversible; what is excluded remains accessible as potentiality. As process, however, the irreversible aspect of time enters the picture. One chooses and then one can never return to the same spot from which the original choice was made, but must continue to choose in a way that is conditioned by previous choices. See, for instance, pages 41–45; for a further complication of these themes, see the discussion of expectations in chapter 8 ("Structure and Time"). The image of the doors marked male and female comes from Lacan (151).

Clearly, similar examples from other areas can be found: a rational theory of the irrational, for instance, or the inner-worldly rejection of the world. For further discussion, see the essays contained in *Reden und Schweigen* by Luhmann and Fuchs.

3. That we could on to infinity is clear, but would not add anything to our analysis of the situation.

4. Here, both Luhmann and Lyotard are in agreement with Derrida. As Richard Beardsworth puts it: "[A] decision is always needed because there is no natural status to language, and that given this irreducibility of a decision, there are different kinds of decisions—those that recognize their legislative and executive force and those which hide it under some claim to naturality *qua* 'theory' or 'objective science'" (12). Theory or objective science here would be equivalent to synthesis, i.e., theory seen as the resolution of, and not as a participant in, a differend.

5. For a fuller discussion of "the jews," see Lyotard, *Heidegger and "the jews,"* especially 3–48.

6. See Derrida's response to Lyotard, which contains the following: "He [Lyotard] lays this inexplicability of Auschwitz (which ought at the very least to invite brevity) to the account of 'Verdrängung,' the 'originary repression,' of which it would serve, in sum, as an example or particular instance. This can leave some perplexity, and says nothing at all about that singularity, if there is one, not to mention those quotation marks around some 'jews.' Who died at Auschwitz, the 'jews,' or some Jews? . . . What is the referent of this proper name, Auschwitz? If, as I suspect, one uses this name metonymically, what is the justification for doing so? And what governs this terrible rhetoric? Within such a metonymy, why this name rather than those of all the other camps and mass exterminations? Why this heedless and also troublesome restriction? As paradoxical as it may seem, respect is due *equally* to *all* singularities" (212).

7. The above reading of Lyotard on "the jews" could be considered a theoretical observation of the practical observation of differend$_1$. What follows, then, could be considered a Lyotardian theoretical observation of differend$_1$. I will leave it to others to offer a practical observation of this theoretical observation. For a more detailed reading of the "ethical" Lyotard from a systems theoretical perspective, see the articles by Stäheli and by Rasch.

8. See the collection of essays and interviews in Luhmann, *Protest.*

9. That a relatively high level of integration or "consensus" occurs within the "negative" space of exclusion, while the "positive" space of inclusion is marked by a relatively low level of integration, is an anti-Habermasian irony that would appeal to Lyotard.

10. Nevertheless, Thorsten Bonacker, following Lyotard here more than Luh-

mann, wishes to identify the logical space that is the other of modernity, the other of communication and rationality, with a realm of feeling and affect that has been constituted by what is excluded in the modern process of subject-formation, and thus leaves the door open for "politicizing" this space as a permanent, if fluctuating, place of resistance. See Bonacker, especially 117–47.

Works Cited

Beardsworth, Richard. *Derrida and the Political.* London: Routledge, 1996.

Bonacker, Thorsten. *Kommunikation zwischen Konsens und Konflikt: Möglichkeiten und Grenzen gesellschaftlicher Rationalität bei Jürgen Habermas und Niklas Luhmann.* Oldenburg: Bibliotheks- und Informationssystem der Universität Oldenburg, 1997.

Derrida, Jacques. "Canons and Metonymies: An Interview with Jacques Derrida." Translated by Richard Rand and Amy Wygant. In *Logomachia: The Conflict of the Faculties,* edited by Richard Rand. Lincoln: University of Nebraska Press, 1992.

Horkheimer, Max, and Theodor W. Adorno. *Dialectic of Enlightenment.* Translated by John Cumming. New York: Seabury, 1972.

Lacan, Jacques. *Ecrits: A Selection.* New York: Dutton, 1977.

Luhmann, Niklas. "Inklusion und Exklusion." In *Soziologische Aufklärung 6: Die Soziologie und der Mensch.* Opladen: Westdeutscher Verlag.

———. *Protest: Systemtheorie und soziale Bewegung.* Frankfurt: Suhrkamp, 1996.

———. *Social Systems.* Translated by John Bednarz Jr. with Dirk Baecker. Stanford: Stanford University Press, 1995.

———. "Speaking and Silence." Translated by Kerstin Behnke. *New German Critique* 61 (1994): 25–37.

Luhmann, Niklas, and Peter Fuchs. *Reden und Schweigen.* Frankfurt: Suhrkamp, 1989.

Lyotard, Jean-François. *The Differend: Phrases in Dispute.* Translated by Georges Van Den Abbeele. Minneapolis: University of Minnesota Press, 1988.

———. *Heidegger and "the jews."* Translated by Andreas Michel and Mark Roberts. Minneapolis: University of Minnesota Press, 1990.

———. *The Inhuman.* Translated by Geoffrey Bennington and Rachel Bowlby. Stanford: Stanford University Press, 1991.

———. *Lessons on the Analytic of the Sublime.* Translated by Elizabeth Rottenberg. Stanford: Stanford University Press, 1994.

———. *Political Writings.* Translated by Bill Readings and Kevin Paul Geiman. Minneapolis: University of Minnesota Press, 1993.

———. *The Postmodern Condition: A Report on Knowledge.* Translated by Geoff

Bennington and Brian Massumi. Minneapolis: University of Minnesota Press, 1984.

Rasch, William. "In Search of the Lyotard Archipelago: or, How to Live with Paradox and Learn to Like It." *New German Critique* 61 (1994): 55–75.

Schmitt, Carl. *The Concept of the Political.* Translated by George Schwab. New Brunswick, N.J.: Rutgers University Press, 1976.

Spencer Brown, George. *Laws of Form.* 2d ed. New York: Dutton, 1971.

Stäheli, Urs. "From Victimology Towards Parasitology: A Systems Theoretical Reading of the Function of Exclusion." *Recherches Sociologiques* 27.2 (1996): 59–80.

Wellmer, Albrecht. *The Persistence of Modernity: Essays on Aesthetics, Ethics, and Postmodernism.* Cambridge: MIT Press, 1991.

10. Blinded Me with Science: Motifs of Observation and Temporality in Lacan and Luhmann

Jonathan Elmer

In taking up the topic of cybernetics in 1955, a field then exerting influence on everything from telecommunications to public health management (see Heims), Jacques Lacan proposed the rubric of "conjectural sciences" for all those sciences of combination, where "[w]hat's at issue is the place, and what does or doesn't come to fill it, something then which is strictly equivalent to its own inexistence" (*Seminar II* 299). This "science of the combination of places as such" is, to be sure, distinct from the exact sciences, which always focus on "what is found at the same place" (299). The exact sciences, in other words, deal with positivities, the conjectural sciences with probabilities. It is, indeed, to Pascal's arithmetic triangle that Lacan turns when he wishes to trace the origins of this science of combinations: "If this is how we locate cybernetics, we will easily find its ancestors, Condorcet, for instance, with his theory of votes and coalitions, of *parties,* as he says, and further back again Pascal, who would be its father, and its true point of origin" (296).

Several years earlier, in *Cybernetics* (a book with which Lacan was familiar) mathematician Norbert Wiener had also reached back to the seventeenth century in tracing the genealogy of cybernetics:

> If I had to choose a patron saint for cybernetics out of the history of science, I should have to choose Leibniz. The philosophy of Leibniz centers about two closely related concepts—that of a universal symbolism and that of a calculus of reasoning. From these are descended the mathematical notation and the symbolic logic of the present day. . . . Indeed, Leibniz, like his predecessor Pascal, was interested in the construction of computing machines in the metal. It is therefore not in the least surprising that the

same intellectual impulse which has led to the development of mathematical logic has at the same time led to the ideal or actual mechanization of
processes of thought. (12)

For Wiener, the development of cybernetics is a story of the power of formal and mathematical reason to introduce order and prediction into
phenomena that do not behave in accord with Newtonian mechanics
and temporality, that are not, as Lacan would say, ultimately always only
where they are, "found at the same place" (see Wiener 30–44). Lacan would
agree with Wiener that the two sciences "are inseparable from one another" (*Seminar II* 296) insofar as both rely on "the little letters" or pure
signifiers that consititute an "ordered register" (299) of signification—
the symbolic—that exists independently from, but renders cognitively
accessible, the real.

But Lacan is, not surprisingly, more interested finally in the differences between conjectural and exact sciences. The central difference
turns on the object of the inquiry, rather than the method: as his allusion
to Condorcet's theory of votes and coalitions indicates, the conjectural
sciences are fundamentally concerned with man—indeed, the rubric itself is meant to substitute for the "group of sciences normally designated
by the term human sciences" (296). In this science of combinations, what
"does or doesn't come to fill" its place is, at bottom, the human being, in
all its various social and psychological itineraries. The terminological
shift from "human science" to "conjectural science" would seem a purposive turn away from the subjective fascinations with human individuality: in making human beings subordinate to the "place" they either do
or do not come to fill, the conjectural sciences make man appear always
under erasure, as it were, alternately materializing and dematerializing,
"strictly equivalent to [his] own inexistence" (299).

The importance of what Lacan thus characterizes as the "conjectural
sciences" in the development of the social sciences can hardly be denied.
Ian Hacking's *The Taming of Chance,* for example, demonstrates the
crucial role, beginning in the eighteenth century, of probability and sta-
tistics in the management of social life.[1] In a slightly different register,
Christopher Herbert has rooted the "ethnographic imagination" in a
readiness to conceive of social reality as essentially relational, a matter
less of positive entities, be they individuals or classes, than of combinations and patterns: what he calls the "culture concept," a "need to think of
culture (in the sense of a complex whole) as the composite of *relationships* existing among the phenomena of a given society" that goes back to

at least the eighteenth century (10). Herbert provides a very interesting account of the transition from moral philosophy to political economy, as exemplified in the career of Adam Smith. What makes this transition so fluid is that both moral philosophy and political economy are essentially what Lacan calls "sciences of the combinations of the scanned encounter" (*Seminar II* 300): that is, whether they deal with the complex sources of envy or deference, or the mysterious beneficence of economic competition, both inquiries extrapolate from an intersubjective matrix in which every calculation must include within it the potential calculations of others. Thus it is that attempts by the social sciences to move beyond the dimension of the merely subjective or intersubjective seem nonetheless to recur, intentionally or not, to the language and imagery of the human encounter: will the other show up or not? Lacan touches on this almost uncanny aspect of "man's waiting": "In the game of chance ... [man] has the idea that something is revealed there, which belongs to him, and, I would say, all the more so given that no one is confronting him" (300). The conjectural sciences emerge as the precipitate from a waiting game.

Sociologist Niklas Luhmann would describe the emergence of conjectural sciences out of the paradigm of the intersubjective encounter as the evolutionary drift of functional differentiation, and more specifically, as the differentiation of social systems and interaction systems. Sometime in the second half of the eighteenth century, it no longer became possible to understand society as "interaction," as essentially a series of face-to-face encounters. Competition for Luhmann, to take a prominent example, "is a non-interactional way of relating to others" ("Evolutionary Differentiation" 118); that is, it expresses the idea that social relations continue to operate even without the immediate presence of others, and hence without all the codified gestures of *politesse* and deference such presence normally requires. The economy becomes less "dependent upon rules of interaction," requiring instead an "understanding of [its] own structural conditions" (117). Economic thought becomes an abstract tracing of the movement of probable combinations. At the same time, however, this evolution toward a "structural understanding" of society was intimately linked to the vicissitudes of face-to-face "interaction": throughout the seventeenth and eighteenth centuries a "new intensity of *social reflexivity* can be observed, of 'taking the role of the other'" (121). In this way, the "interaction system develops a combinatorial space of immense complexity," which itself requires, as it were, the introduction of noninteractional rules—"structural conditions"—to be managed. It is *because* "no participant can know the state of a 'simple' two-person inter-action system" (121), *because* of the phenomenological impasse of what Luhmann calls, following Parsons, "double contingency" (a phenomenon best exemplified historically by the endlessly self-defeating teaching of "sincerity" in eighteenth-century moral discourse) that interaction leads to, as its own surpassing or negation, a structural understanding of socie-ty. As Luhmann remarks, "[t]he awareness of double contingencies auto-catalyzes the development of social systems" (121). Social science, then, whether one understands it as the elaboration of "conjectural sciences" or as the result of the differentiation between society and interaction, finds itself, both historically and conceptually, in a kind of antagonism with a subjective or intersubjective substrate.

This antagonism—or more gently, and less anthropomorphically, this oscillation—between a formalized "structural understanding" of the social and its subjective or intersubjective substrate received its most thoroughgoing consideration thirty years ago, in Michel Foucault's *The Order of Things*. While acknowledging the "double advance" of modern thought, "on the one hand towards formalism in thought and on the other towards the discovery of the unconscious—towards Russell and Freud" (or toward Wiener and Lacan, we might add), Foucault also in-sists on a kind of internal asymmetry to this coupling, an asymmetry glossed over in Lacan's and Wiener's genealogies. For while formalization as an ideal and a method may well link figures like Russell and Leibniz, any consideration of their respective thinking about man is fundamen-tally skewed, suggests Foucault, if it does not recognize the appearance of the "positivity" of man toward the start of the nineteenth century. Thus, while it is "of interest historically to know how Condorcet was able to apply the calculation of probabilities to politics" or "how contemporary psychologists make use of information theory in order to understand the phenomena of learning," such accounts of the thinking about man and his behavior must be tied to the historical recognition that the very ap-pearance of man as an object of inquiry was made possible not by the ex-tension but rather by the "retreat of the mathesis" (Foucault 349). Rather than imagine a mere continuity with the Leibnizian project of a universal symbolic language, Foucault suggests we view the human sciences as bounded on one surface only by the project of formalization. The human sciences do, indeed, "have the more or less deferred, but constant, aim of giving themselves, or in any case of utilizing . . . a mathematical formali-zation." But they also touch on another edge the great empirical sciences Foucault unearths archaeologically: "they proceed in accordance with models and concepts borrowed from biology, economics, the sciences of language." Most importantly, however, the human sciences share a third surface with a philosophical problematic of finitude inasmuch as "they address themselves to that being of man which philosophy is attempting to conceive at the level of radical finitude" (348).

It is this last element in Foucault's extraordinarily intricate archeaology of the human sciences that may be said to be privileged in Foucault's own account. "Modern man . . . is possible only as a figuration of finitude. Modern culture can conceive of man because it conceives of the finite on the basis of itself" (318). What this means is that "man," as an object of inquiry, as an epistemological "positivity," first becomes visible as limited, contingent, and partial. To be sure, there was thinking about human beings in earlier epochs, and there was a kind of Olympian vantage assumed by the classical episteme, a vantage from which all might be surveyed in representation, but that vantage was not theorized as the particular property of man. Indeed, man's very identity, the very thought of his specificity, only emerges, argues Foucault, out of a shift to what we would now call—after Saussure's signifiers have made it so famous—a regime of pure differentiality: "It is apparent how modern reflection, as soon as the first shoot of this analytic [of finitude] appears, by-passes the display of representation, together with its culmination in the form of a table as ordered by Classical knowledge, and moves towards a certain thought of the Same—in which Difference is the Same thing as Identity" (315). Man, and with him the human sciences, emerges not as the peak of a pyramidal hierarchy of being and knowledge, but rather as the very locus of the difference between being and knowledge, the fissure and finitude from which issue all things—knowledge, man himself—in their varied and incommensurable identities.

Contrary then to the caricature of "enlightenment humanism," in which it is taken as an article of faith that optical metaphors figure man's self-appointed sovereignty of the world, in "our humanism"—but what could the earlier humanisms *be*, that had no concept of man?—in "our humanism" man is divided against himself and all else besides: "Man, in the analytic of finitude, is a strange empirico-transcendental doublet," at once a mere finite being, a natural part of a natural world, *and* the "being such that knowledge will be attained in him of what renders all knowledge possible" (318). Man's vision comes into existence with his blindness: "This obscure space"—what Foucault calls the "unthought" to the modern cogito—"is both exterior to him and indispensable to him: in one sense the shadow cast by man as he emerged in the field of knowledge; in another, the blind stain by which it is possible to know him" (326). It is the simultaneous emergence of ignorance and knowledge, blindness and vision, cogito and "unthought" that joins together, at their deepest archaeological strata, the "two great forms of analysis of our day," namely, "interpretation and formalism" (299).

Foucault pursues this idea of man, and the human sciences, as "figuration of finitude" along several trajectories, two of which are important for our present purpose. The first concerns observational metaphors. When man is subject to the great "archaeological mutation" in which he appears "in his ambiguous position as an object of knowledge and as a subject that knows," it does not mean that we simply supersede the infamous classical optic. Rather, the very space of observation is volatilized, internally fissured by the finitude introduced and figured by man. As a result, we get figures of instability and conceptual reversibility, an "enslaved sovereign," and "observed spectator" (312). The second trajectory concerns temporality. If man now comes to stand for an essential instability in the spatial register, both observer and observed, he similarly finds himself figuring the essential instability of historicity: "It is no longer origin that gives rise to historicity; it is historicity that, in its very fabric, makes possible the necessity of an origin that must be both internal and foreign to it" (329). Man becomes the very point of this paradoxical articulation: "the original in man is that which articulates him from the very outset upon something other than himself . . . it is that which, by binding him to multiple, intersecting, mutually irreducible chronologies, scatters him through time and pinions him at the center of the duration of things" (331). Just as man's very emergence into visibility will be figured equally by the triumph of observation and the "blind stain by which it is possible to know him," so too man will be at once the "origin" and "center" of a newly powerful historicity and that same origin's paradoxical temporal implosion, its "recession to itself" (372).

Lest we seem to have strayed too far from the historical dimension outlined by Luhmann and others, let us recall that the disturbances of vision and chronology described by Foucault are the natural outcome of the evolutionary drift away from simple "interaction." For Foucault, too, the crucial moment concerns the infolding of others' calculations in one's own, the doubling-back of representations of the social within the realm of social action:

> there will be no science of man unless we examine the way in which individuals or groups represent to themselves the partners with whom they

produce or exchange, the mode in which they clarify or ignore or mask this function and the position they occupy in it, the manner in which they represent to themselves the society in which it takes place, the way in which they feel themselves integrated with it or isolated from it, dependent, subject, or free. (352–53)

The human sciences must take as their proper object social behavior in which the theoretical plane of representations is inextricably folded into the plane of action itself. For this reason, the thought of the social sciences, just like the social thought they study, is "no longer theoretical." As soon as such thought operates, "it offends or reconciles, attracts or repels, breaks, dissociates, unites or reunites. . . . Thought . . . is in itself an action—a perilous act" (328). Faced with this dilemma of being strung between formalization and intepretation, it is perhaps not surprising that the human sciences have become the terrain for the most vigorous epistemological debate and innovation in the past few generations. Toward the end of *The Order of Things,* Foucault suggests that the insistent attention to their own self-division, their own problem of finitude, has led the human sciences into what he calls an "'ana-' or 'hypo-epistemological' position": "the human sciences, when they duplicate themselves, are directed not at the establishment of a formalized discourse: on the contrary, they thrust man, whom they take as their object in the area of finitude, relativity, and perspective, down into the area of the endless erosion of time" (355). Foucault's analysis leads to a vision of the human sciences as a kind of epistemological sinkhole: unable to pass beyond an analytic of finitude but equally unable to look away, as if were we not to observe man's blind spots, there would be nothing left to observe.

In the analysis that follows, I want to unpack this idea that the human sciences occupy a "hypo-epistemological position." If we look at the grand theoretical endeavors of Lacan and Luhmann, we can see certain shared patterns of figuration, despite other very considerable, indeed irreconcilable, differences in their basic assumptions. Both Lacan and Luhmann are markedly influenced by the "linguistic turn" of the twentieth century, and more specifically, by the promise of formalization seemingly offered by cybernetics' understanding of communication. The psychoanalyst and the sociologist must both contend, however, with the folding-over of theoretical reflection into the object of their inquiry, with the fact that, in Claude Lévi-Strauss's words, theirs are analytic endeavors in which "the observer himself is part of his observation" (29). Both Lacan and Luhmann thus start from that opening wedge between the subject and society, between "interaction" and a structural or formal understanding of social behavior; in this regard, their considerable attention to epistemological matters is dedicated to keeping these two halves from collapsing into one another. But their epistemology is self-consuming, as it were; it is not pursued to clear the field of confusions but rather occupies the entire theoretical field itself. It does this only because it is itself incomplete, with a hole in its middle. At those moments when the theoretical work of these thinkers becomes hypo-epistemological, when it confronts its own internal limit, we see reemerge the disturbances of vision and chronology that lie at the heart of the experience of finitude. Thus, these anti- or posthumanist endeavors reinstate the figure of man in the place of his erasure; man becomes, as Lacan says, "strictly equivalent to his own inexistence," or in Foucault's famous image, a "face drawn in sand at the edge of the sea" (Foucault 387).

I. Lacan and the Blind Gaze

As Foucault remarks at the end of *The Order of Things,* psychoanalysis and ethnology "have been constituted in confrontation, in a fundamental correlation" (379). They are, as it were, inversions of each other, in disputation over how to understand the "double articulation of the history of individuals upon the unconscious of culture, and of the historicity of those cultures upon the unconscious of individuals" (379). In a series of influential texts of the late forties and early fifties—the final pages of *The Elementary Structures of Kinship* (1949), the essay on "The Effectiveness of Symbols" (1949), the *Introduction to the Work of Marcel Mauss* (1950)— Lévi-Strauss entered this confrontation with psychoanalysis, seeking to redirect inquiry away from the "American psycho-sociological school" exemplified by Ruth Benedict and Margaret Mead, and toward his emerging structural anthropology (11). In making this argument in the "Introduction," Lévi-Strauss turned to a new conception of the operation of communication. To be sure, "the ethnological problem is a problem of communication" (36). But in approaching this problem one must look behind the pasteboard egos who communicate, and toward the unconscious of the system, communication itself. Such a perspective no longer discloses intersubjective encounters of a more or less frustrating sort, but the "unconscious itineraries of that encounter" (36), that is, mere communicative functions producing difference in their endless turns. The seemingly insurmountable gulf of difference between subjectivities is surpassed by being functionalized; difference is not a fact of being but the fundamental operation of a system that surpasses any freeze-frame picture of its workings. This interpretive swerve leads Lévi-Strauss to invoke all the motifs so familiar to our structuralist, poststructuralist, systems-theoretical landscape. He encourages us to follow the example of Trubetskoy and Jakobson's structural linguistics, and look for the "infrastructure simpler than any [phenomenological given], to which the given owes its whole reality" (41). Mauss, too, is celebrated for construing the "notion of *function* following the example of algebra, implying, that is, that social values are knowable *as functions of* one another" (43). Ultimately, Lévi-Strauss hopes—as Lacan will ever afterward—for the "progressive mathematisation of the field" (43) of the human sciences, a hope inspired in him by the success of cybernetics' "application of mathematical reasoning to the study of phenomena of communication" (44). Lévi-Strauss theorizes the entirety of what transpires in communication as the workings of a social "symbolic."

In part through the influence of Lévi-Strauss's interest in linguistics and cybernetics, Lacan began in the 1950s to articulate with more precision the relation so central to Lévi-Strauss's theorization of the elementary laws of the symbolic, namely, the relation of the individual to the larger network of social signification in which he or she is caught up. For Lacan, this takes the form of demarcating the respective dimensions of the imaginary and what he calls, with Lévi-Strauss, the "symbolic." To gain some sense of the context in which Lacan began elaborating this distinction so central to all his later thinking, one need only look carefully at his second seminar of 1954–55. Taking up the topic of "The Ego in the Theory of Freud and in the Technique of Psychoanalysis," the seminar was conducted parallel to a series of lectures on "Psychoanalysis and the Human Sciences," which featured presentations by (among others) Alexandre Koyré, Lévi-Strauss (on "Kinship versus the Family"), Merleau-Ponty, Benveniste, and Hyppolite, and which concluded with a lecture by Lacan himself, titled "Psychoanalysis and Cybernetics, or on the Nature of Language."[2] In focusing on the ego, Lacan is, as always, actually hammering home the necessity of moving beyond it; he is continuing his bitter quarrel with the ego-psychology then in power in the International Psychoanalytical Association. Like Lévi-Strauss, Lacan is interested in this seminar in articulating and defining theoretically a trans-subjective "symbolic order" that reflects the new thinking about the formal autonomy of language and communication. But Lacan differs from Lévi-Strauss in taking as his point of departure the inadequacy of the individual's regulative function in relation to the symbolic order:

Beyond the homeostases of the ego, there exists a dimension, another current, another necessity, whose plane must be differentiated. This compulsion to return to something which has been excluded by the subject, or which never entered into it, the *Verdrängt*, the repressed, we cannot bring it back within the pleasure principle. If the ego as such rediscovers and recognises itself, it is because there is a beyond to the *ego*, an unconscious, a subject which speaks, unknown to the subject. (171)

Both Lévi-Strauss and Lacan appeal to a notion of the unconscious that seems to offer a way out of the phenomenological impasses of intersubjectivity. For Lacan, however, there is no way beyond such impasses *but through;* that is, the "beyond of the ego" is apprehensible only because "the ego rediscovers and recognises itself." The turns of reflection and self-consciousness are what reveal the fundamental asymmetry of the individual's relation to the larger social-symbolic world of communication in which he finds himself caught. The unconscious operating "beyond the ego" is tracked in this seminar as a "signifying insistence" (a Lacanian turn on Freud's repetition compulsion), a kind of acephalic desire that disturbs and disrupts the psychic system's fundamental inertia, its homeostatic or restitutive function (60–61). On the other hand, the "insistence" of signification is insistent only for an ego. As Lacan demonstrates in his treatment of cybernetics, one can construct a machine that "embodies the most radical symbolic activity of man" (74), namely, the creation of a "world of symbols . . . organised around the correlation of absence and presence" (300), but one will not for all that have embodied anything like an unconscious. The unconscious, as Lacan develops the concept in this seminar, names the principle of disjunction between the ego and the symbolic, and is "proper" neither to one nor the other. The fact that through "cybernetics, the symbol is embodied in an apparatus . . . in a literally trans-subjective way" (304) does not lead to the conclusion that humans are in essence symbolic, and only epiphenomenally subjective— or that, as Lévi-Strauss remarks a number of times in his essay on Mauss, the symbolic system is "more real" than what carries out its functions. On the contrary, what cybernetics' embodiment "of an order which subsists in its rigour, independently of all subjectivity" (304) effectively makes visible is the *rift* opened up, *within subjectivity as within sociality,* between the imaginary and symbolic registers. This rift is the location of an unconscious that exhibits, paradoxically, both the "radical difference" between the symbolic and the imaginary registers, and the fact that "there is something in the symbolic function of human discourse that cannot be eliminated, and that is the role played in it by the imaginary" (306). Imaginary and symbolic require each other for their own articulation, but only articulate themselves as disarticulated from each other.

How does Lacan arrive at such a complex and paradoxical notion? The fact that there exists a "role played in it by the imaginary" that "cannot be eliminated" may well seem troubling to the notion of the autonomy of symbolic signification. If the symbolic order is only ever revealed—experientially, theoretically, or analytically—via the imaginary, what prevents the very autonomy of the symbolic from being simply another version of imaginary misrecognition? After all, it is precisely the function and role of the imaginary to secure the illusions of autonomy and identity out of a fundamental alienation. Lacan confronts this question directly toward the end of his lecture on cybernetics and psychoanalysis: "The issue is to know whether the symbolic exists as such, or whether the symbolic is simply the fantasy of the second degree of the imaginary coaptations" (306–7). He will insist in this seminar, and with increasing vehemence throughout his later work, that the symbolic does indeed "function in the real, independently of any subjectivity," because the option for him is to imagine the possibility of reducing this symbolic "beyond" to the functional dimensions of the imaginary, of dissipating its transcendent status through an ego-psychological "rectification," or "normalisation in terms of the imaginary" (307)—in other words, to capitulate to the ego-psychologists' goal of helping the patient produce a "healthy ego." But Lacan's insistence on the simultaneous inextricability and irreducibility between imaginary and symbolic registers means that he will commit himself fundamentally to observational scenarios, in which the move "beyond" the imaginary looks more like an internal crack or failure within imaginary reflection itself.

It is thus not surprising that Lacan develops his most famous parable of observation, his reading of Poe's "The Purloined Letter," in this theoretical context. In both its full text in *Ecrits,* and in its initial development in the course of the seminar, the treatment of "The Purloined Letter" concerned the way in which the symbolic "is attained most especially on the imaginary level" (177). What Poe's story stages for Lacan, both in its essential plotting concerning the letter's displacements and, just as crucially, in the interpolated commentary about the game of even and odd, is the recognition that purely imaginary observational scenarios have *built into them* a kind of essential fracture, a place of failure or opening, that opens onto the symbolic: "What's at issue is an essential alien *[dissemblable],* who is neither the supplement, nor the complement of the fellow being *[semblable]*, who is the very image of dislocation, of the essential tearing apart of the subject. The subject passes beyond this glass in which he always sees, entangled, his own image" (177). A page later Lacan repeats: "Under certain conditions, this imaginary relation itself reaches its own limit, and the ego fades away, dissipates, becomes disorganized, dissolves" (178). In "The Purloined Letter" the series of observers of observation (the Queen, the Minister, Dupin, etc.) believe themselves to have seen the dynamic of looks in its entirety—to have, as it were, the imaginary, intersubjective scenario fully laid out before them. The observers of observation are not wrong; they correctly assess what they see in the room. Lacan's point is rather that we have here a parable of the imaginary relation reaching its own point of dissolution, in which the observational vantage cannot stay merely observational, but finds itself caught up, precipitated, into a symbolic circuit of action. It is this position in the symbolic circuit, defined by the relation to the letter, that most profoundly confers identity on the story's players, and not any one character's adequate or inadequate reflective understanding.

The appearance of the symbolic dimension beyond the imaginary is, then, a problem of action and of time. To demonstrate this more clearly, we will refer to an earlier essay, the essential elements of which reappear at the end of the 1955 seminar, where the essay is described as illustrating precisely the importance of "distinguishing the imaginary from the symbolic" (287). It is a text originally written in 1945, titled "Logical Time and the Assertion of Anticipated Certitude: A New Sophism." I will quote John Forrester's translation of the scenario, which can be found in his excellent commentary on Lacan's theories of temporality. A "prison governor" must free one of three prisoners, and will allow their "lot to be determined by a test":

> There are three of you here. Here are five discs which differ only in their colour: three are white, and two are black. Without letting you know which of them I have chosen, I am going to fasten one of these discs between each of your shoulders, that is to say, out of the bearer's direct vision; all possibility of his being able to catch sight of it indirectly is also excluded by the absence here of any means of looking at himself.
>
> Thereafter you will be free to consider at your leisure your companions and their respective discs, without being allowed, of course, to communicate to each other the fruits of your inspection. . . . [T]he first who can deduce his own colour shall profit from the measure of liberty of which we dispose. (Forrester 178–79)

All three prisoners are then given white discs. "After they have pondered a certain time, the three subjects together make a few steps toward the door, arriving there abreast." They each give the following reasoning:

> I am a white, here is how I know it. Given that my companions were whites, I thought that, if I were a black, each of them would have been able to make the following inference: "If I were also a black, the other, immediately realising from this that he is a white, would have left straight away; therefore I am not a black." And the two others would have left together, convinced of being whites. If they stayed put, it is because I am a white like them. (179)

Forrester asserts that this sophism serves Lacan in his later work as "*the model for all the relations between the subject, a set of signifiers, and temporality*" (179–80). From our perspective, its importance lies in its presentation of the modulation, through a distinctive temporality, between imaginary and symbolic registers. The strictly reflective, imaginary methods at the disposal of the prisoners can lead to a knowledge of the symbolic structure in which they are caught only by virtue of an odd temporal jump between what Lacan calls the "time for understanding"—"directly articulated upon the time of meditation of the other," as Forrester writes— and what he calls the "moment for concluding." Out of a strict intersubjectivity, in other words, emerges a dimension beyond, one in which certitude cannot be had through a mere reflective assessement, but can only be retroactively verified by an action that is already presented as being *too late*: "From hesitating, a pulsation leads immediately to being too late. This precipitation is not simply a contingent effect of the dramatic situation; the subject must make haste, because if he does not, and the two others beat him to it, then he will no longer be sure that he is not black" (181).

Commenting on this sophism ten years later in the seminar, Lacan claims that it demonstrates that a "relation to time peculiar to the human being" governs this movement of haste beyond the intersubjective: "there is a third dimension of time which [cybernetic machines] are not party to . . . which is neither belatedness, nor being in advance, but haste. . . . That is where speech is to be found, and where language, which has all the time in the world, is not" (*Seminar II* 291). And in the lecture on cybernetics, he makes the point again: "With a machine, whatever doesn't come on time simply falls by the wayside and makes no claims on anything. This is not true for man, the scansion is alive, and whatever doesn't come on time remains in suspense. That is what is involved in repression" (307–8).

When Lacan writes, in the seminar, that "the subject [of the sophism] holds in his hands the very articulation by which the truth he sifts out is inseparable from the very action which attests to it" (289) we have in fact a double, paradoxical articulation. The subject can "know" something only by acting in haste, by a "precipitation in the act" (289), a precipitation that is always ahead of the verification it brings. If it is in some sense true that the subject "holds in his hands" the articulation necessary to arrive at the truth, then, it could just as easily be said that the subject's "precipitation in the act" just *is* the articulation between the truth arrived at in the imaginary "time for comprehending" and its manifestation in the symbolic "moment for concluding." This temporal lurch—like stepping onto the moving platform of a carousel—is that whereby the subject modulates, perhaps agonizingly, but always in the anxiety of "haste," between he who articulates signifiers, and he who merely embodies the articulation of signifiers. The vantage from which it seems that the subject manipulates signifiers (white, black) to represent another subject (himself), suddenly flips over into a temporalized movement in which the "signifier . . . represents the subject for another signifier" ("Subversion of the Subject" 317).

Lacan here imagines man at a point of radical temporal disjunction, a "precipitation in the act," a tumbling over or lurching movement. Man does not articulate here, but is shown to be articulated temporally. Thus, the "precipitation in the act"—even as it, in anticipating, moves beyond whatever certitude the reflective ego can attain, even as it presents the surpassing of the self—does not move beyond identity but rather bores into its most secret mechanism. A seemingly semantic pattern can help draw out this connection between imaginary and symbolic identity. In the essay on "Logical Time," Lacan describes a process of "decanting" by which the "I"—the *"je"* of the conclusive assertion (which is presented definitively only in the act of making for the door)—is drawn off from the earlier moments of reflection, each with their appropriate "subject" forms.[3] This final identity-marker, the one bound up with the anticipated certitude of the act, is

> isolated by a beat of logical time with the other, that is with the relation of reciprocity. This movement of the logical genesis of the *je,* by means of a decanting of its own time, is parallel to its psychological birth. In the same way that . . . the psychological "I" disengages itself from an undetermined specular transitivism . . . the "I" in question here is defined by the subjectivizing of a competition with the other in the function of logical time. (208)

Given this explicit parallel between the "logical genesis" of the *je* and its "psychological birth," it is perhaps not surprising that the same odd figure of "decanting" should emerge later in Lacan's treatment of the construction of identity-in-difference that is the mirror stage. Samuel Weber has directed attention to what he takes to be a crucial addition to the theorization of the mirror stage, a revision articulated by Lacan in his paper on Daniel Lagache's "Psychoanalysis and the Structure of the Personality." "The context," writes Weber, "is defined by the question of the Other (capital O): that is, by the function of alterity or heterogeneity in discourse" (116). (It is, of course, precisely this question that animates the entirety of the second seminar. Jacques-Alain Miller draws attention to this trajectory by titling the third section of the text "Beyond the Imaginary, the Symbolic; or, From the Little to the Big Other.") Weber notes a slight revision in the canonical presentation of the child's recognition and jubilation before the mirror:

> In contrast to Lagache's "personalistic" interpretation of Freudian doctrine, Lacan stresses the impersonal structure "of this Other, where discourse is situated"; such alterity, he continues, reaches to "the purest moment of the mirror relation." What is this "purest moment"? Lacan locates it [and here Weber translates the passage to be found in *Ecrits* (678)] in the gesture by which the child at the mirror, turning around to the person carrying it, appeals with a look to the witness who decants, by verifying it, the recognition of the image from the jubilant assumption, in which, to be sure, *it* [such recognition] *already was*. (116)

Let us clarify what is at issue here. Weber understands the "verification" being provided by the "witness" as intervening between two moments— recognition and jubilation—that had earlier seemed coeval:

> In the original version of the text of the mirror stage, the effect of the reflection upon the child seemed to result from its recognition of the image as its own likeness. . . . In the later essay, recognition is no longer enough; instead, in its stead, there appears the anxiety which causes the child to twist back, turning around, and in this gesture, to seek the confirming look of another. . . . [T]he jubilant reaction does not relieve the child from having to seek something like an acknowledgement of the other. In this sense, recognition is no longer a process organized around two poles: child and mirror image, subject and object. Instead, it emerges as a triadic relation in which the acknowledgement emanates not from the self-identical ego, but from the "person who carries it," that is, from the place of the Other. (Weber 118)

We could say, perhaps, that the jubilant assumption of identity in the mirror stage, in the imaginary, was a bit hasty, a bit precipitate, since it now seems to need to be verified. But recall that the lurch into the symbolic in the essay on logical time was exactly *coincident with* this precipitation into action, and that this was also the verification of the truth that could be certified in no other way for the merely reflective ego. The temporal suspension at work in this imbrication of the symbolic within the imaginary construction of identity is once again on Lacan's mind. The "event" in question here is the very assumption and recognition of identity, whose status is forever suspended in a temporal distention by the "tertiary presence" of the symbolic.

But what about this mysterious "witness" who appears in this essay, who verifies, who decants, and who, as Lacan insists, "owes nothing to the anecdotal figure who incarnates it" (*Ecrits* 678)? Weber remarks that this Other "can in essence be determined neither as an individual, nor as a social function, nor as a subject in general. Indeed, it is nothing more than the differentiality upon which discourse depends" (119). The status of the "witness" as pure differentiality would seem to accord well to its temporal paradoxicality, its coming to be only in no longer being there. This mysterious Lacanian "witness" who stands behind the games of mirror-identity and ego-construction is precisely the guarantor of that game in constituting its exception and undoing. This is a witness who verifies by disappearing, whose temporal fading as pure differentiality attests to a symbolic order that is beyond time. Such would seem to be the implication of a particularly metaphysical moment toward the end of the second seminar:

> The wager lies at the heart of any radical question bearing on symbolic thought. Everything comes back to *to be or not to be,* to the choice between what will or won't come out, to the primordial couple of *plus* or *minus.* But presence as absence connotes possible absence or presence. As soon as the subject comes himself to be, he owes it to a certain non-being on which he raises his being. And if he isn't, if he isn't something, he obviously bears witness to some kind of absence, but he will always remain purveyor of this absence, I mean that he will bear the burden of its proof for lack of being capable of proving the presence. (192)

There is a structure of debt here, and of witnessing, which issues ineluctably from the "primordial couple of plus or minus," presence or absence, being or non-being. But this structure is asymmetrical, out of whack; the second terms—minus, absence, non-being, in short, differentiality—always have the upper hand. If the subject comes to be, "he *owes* it to a certain non-being on which he raises his being." In notable contrast, the "structure of that tertiary presence" of the witness "*owes nothing* to the anecdotal figure that incarnates it." This witness guarantees, stands surety for, the game that plays out the inevitable collapse of all witnessing. In the "wager" that starts with pure differentiality—the wager of a postontological social thought founded on the nonfoundation of a signifying order with no positive terms—the game is secured, the strucural order guaranteed and verified, by the very collapse or evanescence of the terms or figures who serve as its "human support" (192). Negation is primal, since the "primordial couple" is such only through its own self-canceling.

At the same time, however, we may well question this erection of a debt-free symbolic realm, one that owes nothing to its human supports, its only theater. For, as Lacan himself insists, this order of nonbeing shares one essential feature with its human support: an insistent desire. Here are the closing words of the seminar: "The symbolic order is simultaneously non-being and insisting to be, that is what Freud has in mind when he talks about the death instinct as being what is most fundamental—a symbolic order in travail, in the process of coming, insisting on being realised" (326). The paradox of this relation between being and nonbeing is what realizes itself in the strange image of the witness or blind gaze in Lacan's thought. For it is precisely by demanding that the subject "verify" its (self)recognition through an appeal to a "witness" who fades away before one's eyes, and who thus images the "essential tearing apart of the subject"—it is precisely in giving over to the human his own self-superseding in signification that nonbeing comes to be. The big Other, which has no ontological status, paradoxically comes to be, realizes itself, in the movement whereby the human recreates himself in the image of the blind gaze to which, he can't help feeling, he owes everything.

When Foucault writes that the social order and the figure of man are cocreated on the ground of a radical finitude, he could well be describing Lacan's construction of the symbolic. Just as Foucault writes of the "blind stain by which it is possible to know" man, Lacan talks of the symbolic as "like an image in the mirror, but of a different order." This order is a kind of monstrous beyond of the imaginary, a beyond of the elaborate optics defining the positionality of self and other. "It isn't for nothing that Odysseus pierces the eye of the Cyclops," Lacan continues gnomically (185). The symbolic can only appear through a mutation process within the imaginary, in which vision and reflection is first inflated to the monstrous proportions of the singular Cyclopean eye, and then blinded, passed beyond, negated. Such a negation, such a recognition of the stain in the mirror, which is also already a passing beyond it, is perhaps the ultimate "precipitation in the act," the temporal lurch bestowing identity. Lacan comments that it is only by blinding the Cyclops, creating of him a blind gaze, that the subject can open communication. This symbolic realm of communication is the home of negation and radical finitude, in which man attains a paradoxical freedom and identity in knowing himself to be, like Odysseus, "No Man."

II. Luhmann and the Constructed Observer

Niklas Luhmann's systems-theoretical sociology does not underestimate the immense creativity of Odysseus's negating prowess.[4] Luhmann's work is a career-long celebration of the fecundating power of distinctions, and reading his work can at times be a dizzying enterprise, as his terms and oppositions reproduce in a process of discursive mitosis. His is also one of the more extended and impressive explorations of the power of systems to operate paradoxically and self-referentially, which makes it hardly surprising that his arguments rarely proceed in a linear fashion, but rather by a reiterative process. Luhmann's *oeuvre* is vast and complex, even for someone who reads him in translation; there can be, therefore, no question of surveying the entire range and import of that work.[5] I want instead merely to present enough of his theoretical edifice to situate an interrogation of a particularly insistent distinction in his theory, namely the unconditional divorce between what he calls psychic systems and social systems.

Like Lévi-Strauss and Lacan, though perhaps with even more vehemence, Luhmann wishes to avoid the lures of any theoretical understanding based on models of the subject of reflection: "The subjectivist problem was to state and to show how it is possible by means of *introspection*—that is by the passage to the self-reference of one's own consciousness—to form judgments about *the world of others*" ("Cognitive Program" 66). It is *not* possible, asserts Luhmann; and indeed the realization of this limit was both the result of, and the further spur for, the differentiation of society and interaction in modernity—that is, the empirical and theoretical divide between a sociality understood on the model of a face-to-face intersubjectivity (which is, after all, finally the model of subjectivity) and one thought of as operating according to its own non-subjective operations. A fundamental consequence of discarding subjectivist approaches to sociality is the recognition, by now canonical, that

there is no vantage—even the theoretical one of "absolute knowledge"—
from which one can perceive and know a reality as absolutely external to
oneself. The optics of self-reflection, inaugurated for modernity in
Cartesian dualism, must be replaced by a functionalized understanding
of primal difference. Thus we get the familiar view of a signifying order
beyond representationalism or realism: "Cognition is neither the copying
nor the mapping nor the representation of an external world in a system.
Cognition is the realization of combinatorial gains on the basis of the
differentiation of a system that is closed off from its environment (but
nonetheless 'contained' in that environment)" ("Cognitive Program" 69).

The notion of a "system closed off from its environment" is fundamental to Luhmann's thought. Following both systems theory and, more
crucially, the theory of autopoiesis of Maturana and Varela, the notion of
closure here put forward needs to be understood in all its equivocalness.
A system effects closure not through a radical indifference to its environment, a pure unaffectedness; rather, being operationally closed means
that changes in the environment can function only as "triggers" for the
system, and take effect in the system only when, and if, they are coded by
the internal system as systemic (and not environmental). Such a view of
the system/environment relation emphasizes autonomy and closure; it
repudiates strict causal interpretations (where, for instance, an environmental factor can unilaterally "cause" a change in a given system) and
it disallows input-output models (wherein a system might be seen to
"process" environmental factors). Most importantly for our argument, it
is fundamentally opposed to any view of the relation between system and
environment based on modeling, or likeness; there can be no ground on
which two systems can recognize themselves as "like each other," since
there is no system that does not reproduce itself save by closure, by differentiation: "The function of the boundaries [of any system] is not to pave
the way out of the system but to secure discontinuity" (66).

Like Lacan's irreducible function of misrecognition, this view of system closure is bracingly antihumanist when considered from the vantage
of any psychology:

> By this means [the theory of autopoietic systemic closure], the significance
> of psychological epistemologies is considerably reduced, but relieved at
> the same time of the unreasonable expectation that they should provide
> more than individual-psychological knowledge. There is no such thing as
> "man," no one has ever seen him and if one is interested in the system of
> observation that organizes its distinctions by means of this word or concept

one discovers the communication-system called society. There are now approximately 5 billion psychological systems. It has to be asked which of these 5 billion is intended when a theory of knowledge employing a psychological reference system relates concepts such as observation and cognition to consciousness. (78)

The same point is made in slightly different terms in a recent article specifically dealing with the relation between psychic and social systems:

> Everyone knows, of course, that the word "human being" is not a human being. We must also learn that there is nothing in the unity of an object that corresponds to the word. Words such as "human being," "soul," "person," "subject," and "individual" are nothing more than what they effect in communication. They are cognitive operators insofar as they enable the calculation of continued communication. . . . The unity that they represent owes its existence to communication. ("How Can the Mind Participate?" 387)

The names for identity do not refer outside the system in which they operate; that is, there is no "real" unity of human identity corresponding to the word "soul," and so forth, there is only what communication can do with this word. The system-reference for such signifiers is not, in other words, the psychic system, but the *social* system, for it is only the social system, according to Luhmann, that reproduces itself by means of communication; the psychic system reproduces itself as an autopoiesis of consciousness ("Autopoiesis" 2). In the communication system that simply is society, "subjects are spoken of to define the self-referential foundation of the cognitions of the mind" ("How Can the Mind Participate?" 387). Consciousness, as an autopoietic system, may have a kind of paradoxical and self-referential unity, but it has nothing to do with the unity of a word like "person," which is merely a "cognitive operator," and thus "owes its existence to communication." In short, Luhmann agrees with Lacan on this question: the signifier represents the subject for another signifier. It is only in order to keep communication going that the unities of "person" and so forth are fabricated. Because psychic systems are wholly environmental for social systems, the latter have to fabricate their own discursive unities—the notorious ideologemes of humanism: "soul," "person," "subject," "individual," and "human being."

But why? Why should communication systems—that is, social systems—need to create this figment of individual identity? This question touches on one of the most difficult issues in Luhmann's theoretical enterprise, the description of the mode of relation between psychic systems and social systems. For it turns out that they are not only, or not merely, environmental for one another, despite Luhmann's insistence that they remain so in the theoretical last instance. Their relation is more complicated. In what Luhmann concedes is a less than satisfying expression, he claims that psychic and social systems "interpenetrate." He explains: "'Interpenetration' does not refer to a comprehensive system of coordination or to an operative process of exchange (something that would require being able to talk about inputs and outputs in this sense). 'Interpenetration' can only mean: the unity and complexity (as opposed to specific conditions and operations) of the one is given a function within the system of the other" ("How Can the Mind Participate?" 386; see also "Individuality" 117). The name for the interpenetration of psychic and social systems when seen from the psychic system is "socialization" (386). "Communications systems," on the other hand, "experience interpenetration by considering the personal dynamics of humans in their physical and mental (including the mind) dimensions" (386). But it is puzzling that a strictly nonpsychic system of communication could be in a position to "consider . . . personal dynamics," much less "experience" anything at all. As with Lacan's mysterious characterization of the symbolic dimension "insisting" on being, we return here, in Luhmann's theory, to a discursive, signifying system that in essence opposes and excludes the psychic dimension, and yet somehow is endowed with expressive desires, insistence, abilities to "experience." What does the communication system's "consideration of personal dynamics" look like?

Answering this question leads us back to the motif we have been examining throughout this essay—namely, the oddly central status of the figure of the observer in theories of nonsubjective signifying orders. In order to comprehend Luhmann's complex answer to this question, we need to widen the perspective somewhat. In an early programmatic essay titled "Meaning as Sociology's Basic Concept," one does not find such emphasis on the radical disjunction between psychic and social systems. The influence of Husserl and Parsons on Luhmann's thinking is more pronounced at this point in his career; and it is Parsons, indeed, who bequeaths to Luhmann the problem of "double contingency": "All experience or action that is oriented to others is doubly contingent in that it does not depend solely on me, but also on the Other, who I must regard as an alter ego, i.e., as just as free and unpredictable as I am" ("Meaning" 45). In this early essay, the transition from such intersubjective scenarios to social structures is considerably more integrated than it will later

become. It is still possible for Luhmann to write that, while "social structures do not take the form of expectations about behavior"—that would
be to return to a notion of sociology as concerned with the elemental
unit of "action" rather than "meaning," precisely the notion being disput-
ed in this essay—they can be understood as the maintenance of a certain
"level of reflexive expectation," or "expectations about expectations"
(45). There would seem, then, to be a crucial role for a kind of intersubjective reflection in the creation of social structures.

At the same time, however, "meaning" is already operating in this early
essay as the conceptual wedge between psyche and sociality, opening up
the possibility, more and more vigorously pursued through the next
twenty years, of emptying out this reflection of any subjective reference.[6]
Habermas has criticized Luhmann's theory as a philosophy of the subject
wholly emptied out of subjectivist depth:

> Systems theory has to remove from the "self" of the relation-to-self all
> connotations of an identity of self-consciousness established by synthetic
> performances. Self-relatedness is characteristic of individual systemic ac-
> complishments in their mode of operation; but no center in which the sys-
> tem as a whole is made present to itself and knows itself in the form of
> self-consciousness issues from these punctual relations-to-self. In this way,
> the concept of reflexivity is separated from consciousness. But then an
> equivalent is needed for the conscious substrate of the self-relatedness that
> is distinctive of the level of sociocultural life. As an emergent attainment
> corresponding to consciousness, Luhmann introduces a peculiar concept
> of "meaning." (Habermas 369)

Whether or not one considers Luhmann's concept of meaning to be "peculiar," it is surely right to recognize that what allows for "the concept of
reflexivity [to be] separated from consciousness" is Luhmann's theorization of meaning, and more specifically, "communication." The theoretical
extrication of reflexivity from consciousness by means of a kind of discourse theory entails, however, as it did for Lacan, the rhetorical *intrication* of the figures of observation with the development of the idea of
communication. What the figure of observation offers, for this paradigm, is the possibility (perhaps even the inevitability) of a vantage on
the operation of the communication system, of seeing the code *as* code.

This helps to explain why one sees observers everywhere in Luhmann's thought. In "The Cognitive Program of Constructivism," his
most thoroughgoing epistemological treatment of this theme, one comes
across sentences that sound like abstract synopses of Poe's "The Purloined

Letter": "Constructivism describes the observation of observation that concentrates on how the observed observer observes. This constructivist turn makes possible a qualitative change, a radical transformation, in the style of recursive observation, since by this means one can also observe what and how an observed observer *is unable* to observe" (73). In part, this predilection for the language of observation is due to Luhmann's incorporation of the vocabulary of second-order cybernetics. At the same time, we should note that Luhmann's desire to ground a notion of reflexivity beyond the subject leads him to emphasize not observation as such, but rather its constitutive *limit*. Luhmann's metaphor of observation is, as it were, a self-canceling figure, and thus very like Lacan's in this regard. For the entire point of observation is that it always proceeds by way of a fundamental blindness: "The 'blind spot' of each observation, the distinction it employs at the moment, is at the same time its guarantee of a world" (70). Later in the same essay he makes the point more fully:

> For cognition, only what serves in a given case as a distinction is a guarantee of reality, an equivalent of reality. One could say more precisely: The source of distinction's guaranteeing reality lies in its own operative unity. It is, however, precisely as this unity that the distinction cannot be observed—except by means of another distinction which then assumes the function of guarantor of reality. Another way of expressing this is to say that the operation emerges simultaneously with the world which as a result remains cognitively unapproachable to the operation. (76)

Every observation involves making a distinction, the background operative unity of which is that observation's enabling blind spot, at once constituting the system's operational closure and bringing forth a world. It would seem that one needs the figure and errancy of observation in order to serve as cover for the more profoundly creative power of blindness.

We can already see, I think, how Luhmann's use of this figure echoes that of Lacan, for we have here an anti-intuitionist, antihumanist account of the construction of meaning systems that constitutes itself on the back of a blinded observer. In reading Luhmann on the difference between psychic and social systems, it often feels that Habermas's characterization is apt: "The flow of official documents among administrative authorities and the monadically encapsulated consciousness of a Robinson Crusoe provide the guiding images for the conceptual uncoupling of the social and psychic systems, according to which the one is supposedly based solely on communication and the other solely on consciousness" (Habermas 378). But Luhmann's account of the way psychic systems participate in communication is more complex than this characterization, for in fact the two systems "interpenetrate" in a mutually interfering way reminiscent of the symbiotic antagonism between Lacan's symbolic and imaginary registers.

Thus, in recent essays, Luhmann has returned to the question of the construction of an observer in the kind of intersubjective scenarios he has been treating with suspicion for some time. Given his own idea of the operational closure of psychic systems, he asks, "why would an observer observe another observer as observer, as another psychical system? Why isn't the other system seen simply as a normal object in the external world, that is, why isn't it simply observed directly instead of as a pathway for the observing of its observing?" ("Cognitive Program" 79). From the perspective of the psychic system, Luhmann is asking, how does it happen that another psychic system can be imagined as not merely an object? Luhmann rejects psychological accounts in which "it is usually assumed that this is made possible by a sudden, intuitive analogy: the other is experienced as an alter ego, as operating like another I" (79); he dismisses as well Maturana's attempt to deal with the issue from a biological perspective. Writing with a not uncharacteristic disciplinary partisanship, Luhmann puts forward a "third theoretical suggestion, (which draws on sociology, since psychology and biology have not sufficed)," namely that "the other observer is a necessary consequence of communication" (79).

What Luhmann describes here is a kind of trickery by which the system of communication creates the illusion for a psychic system (whose "observation" here is indeed neurophysiological, that is, a matter of perception and consciousness) that there is another unity—another observer—with whom it can communicate:

> Within the communication system we call society, it is conventional to assume that humans can communicate. Even clever analysts have been fooled by this convention. It is relatively easy to see that this statement is false and that it can only function as a convention and only within communication. The convention is necessary because communication necessarily addresses its operations to those who are required to continue communication. ("How Can the Mind Participate?" 371)

A communication system needs psychic systems to reproduce itself, but it can only use psychic systems if they are mystified as to their own ability to communicate. Those who are "required" to continue communication are, paradoxically, those who precisely cannot do so, namely the psychic systems called "humans." We get a sense of the stringency of this requirement in Luhmann's characterization of the mind's "participation" in communication: "[I]t is sufficient for the communication process to understand that the mind, virtually helpless, must participate" in the fabrication of this unity of an "other" with whom to communicate. And indeed, in a conception reminiscent of Lacan's elaborate schemas of the optical delusion of the imaginary, Luhmann here relies on the idea of "fascination," presenting the mind the way explorers and conquistadors used to describe the natives, as fascinated by shiny objects: "Language and script fascinate and preoccupy the mind and thereby secure its participation" ("How Can the Mind Participate?" 376); these technologies of communication constitute "special experiential objects that are either extraordinary or fascinating" (375). It is in this way that communication leads the psychic system into a "detour": "The detour via communication, the participation in a completely different operating system, and the *attractiveness of the constitutive difference* of this system are all critical for the constitution of an alter ego" (emphasis added)—that is, for the critical, though illusory, "convention" that there exists an other with whom I can, as psychic system, communicate.

It is not entirely clear what this "attractiveness of the constitutive difference" between psychic and social system would be. Even as Luhmann puts forth an account of how psychic systems are lured into imagining and constructing a sameness, he also sees at work a kind of alluring differentiation. It turns out, indeed, that psychic systems are crucially engaged in distinguishing, and thereby introduce what could be characterized as a rhetorical dimension to the interpenetration of psychic and social systems. "Communication is only possible," writes Luhmann, "when an observer is able, in his sphere of perception, to distinguish between the act of communication and information, that is, to understand communicative acts as the conveying of information (and not simply as behavior)" ("Cognitive Program" 79). What psychic systems are really required to do is distinguish—and that this distinguishing is a matter of the psychic system is emphasized by Luhmann's reference to perception—between constative and performative dimensions of language. Communication cannot continue if it does not reproduce itself by means of the distinction between information and utterance—or, as the same distinction is translated elsewhere, between "facts" and "behaviors." In other words, communication is self-divided; although not itself anything psychic, much less bodily, it nonetheless needs such psyches and bodies to operate, and what it needs them to do is constantly to renew their own distinction from the communication system, by indicating the difference between information and utterance. Communication systems are not merely interpenetrated by psychic systems, they are incomplete without them; but paradoxically they can only effect closure, complete themselves operationally as it were, by having their own incompletion or dividedness rearticulated. Communication reproduces itself by luring psychic systems into attributing certain communications to other psychic systems as their "actions" ("Autopoiesis" 6–7). In doing this, communication both refers to its own constitutive distinction between information and utterance, between hetero-reference and self-reference, *and* it fabricates and presupposes the synthesis of these distinctions in the attribution of action. Humans, psychic systems, are communication's alibis, its constantly rearticulated "presupposition" of synthesis.

This presupposition of synthesis is an operative element that cannot be reduced further. When Luhmann describes the autopoiesis of social systems in communication, he appeals to such an "undecomposable unit" upon which, at every moment, in every linkage or articulation, the system depends. This unit, for communication, is the "synthesis of information, utterance and understanding" ("Autopoiesis" 4), a synthesis that is "operative": "As an operating unit it is undecomposable, doing its autopoietic work only as an element in the system" (4). At the same time, because we are here dealing with a meaning-based system that can "re-enter" its own constitutive distinction from the environment within its system, communication can only "observe" itself, "reenter" or recur, by disarticulating this very distinction:

> It is forced by its own structure to separate and to recombine hetero-referentiality and self-referentiality. Referring to itself, the process has to distinguish information and utterance and to indicate which side of the distinction is supposed to serve as the base for further reference. Therefore, self-reference is nothing but reference to this distinction between hetero-reference and self-reference (4).

That is, communication's self-reference and psychic self-reference (here considered as the perception of the difference between information and utterance considered above) are one and the same, considered from the vantage of their results. Just as much as psychic systems, the self-reference of the communication system only takes place by the insistent disarticulation of selves and reference, bodies and language, or in Luhmann's terms, psychic and social systems. But with each new disarticulation, the fiction of the synthesis of self- and hetero-reference must be "presupposed" anew. The working of self-reference, its essential differentiating movement, requires at every moment a mirage of identity from which to distinguish itself, a synthesis to cancel and disarticulate. Every observation proceeds only on the back of the presupposition of a prior, "blind" operation. (Luhmann asserts unambiguously that "operations of the mind and of communication proceed blindly" ["How Can the Mind Participate?" 382]). The figure of observation requires then, for its very existence, the concept of blind operation. In the "construction of an observer," of the blind, helpless, fascinated gaze of a deluded psychic system, the communication system secures its own operational closure by imagining it has successfully located its own environment. And the same holds for those perceiving psychic systems forever differentiating themselves through reference to the distinction between communicative behaviors and mere information.

What Luhmann's theory allows us to say is that meaning-systems require the alibi of identity in order to keep going. This identity is always presupposed, because always already undone, a synthesis "blind" to its own internal rupture or constitutive difference. In pointing out this rupture and this blindness, "observation" creates the necessary next fiction of "operation." The problem is that this "observation" cannot but presuppose the "operation," and in positing a blinded observer—say, a psychic system deluded about its ability to communicate—the observing system merely regards itself in the mirror without recognizing that fact. Insofar as *both* psychic and social are essentially meaning-based systems with the ability to "reenter" their own constitutive distinction, their "interpenetration" is precisely their simultaneous operation, as each observes its self-distinction and thus presupposes the necessary operation in the other. They are inextricably joined in their inability to proceed without constantly disarticulating one from another, self- from hetero-reference. Their essential unity lies in their paradoxical operational commitment to "closure." Because "identity" is a problem for them, because they cannot achieve identity otherwise than differentially, theirs is a symbiotic antagonism; each functions as the operation for the other's observation, the alibi of an environment for the other's system.

But in a swerve that should no longer be surprising, these observational paradoxes are reformulated, or recoded, by Luhmann as temporal predicaments.[7] The structure I have just been describing, in which the "continuing dissolution of the system becomes a necessary cause of its autopoietic reproduction" ("Autopoeisis" 9), is described in temporal terms when Luhmann writes that "conscious systems and social systems have to produce their own decay," that they must "produce their basic elements . . . not as short-term states but as events that vanish as soon as they appear" (9). These events serve as another version of the "presupposed synthesis" allowing for autopoiesis, and thus retain a maximum of paradoxicality: "Events can be identified and observed, anticipated and remembered only as such a difference. Their identity is difference. Their presence is a copresence of the before and the thereafter. They have to present time within time and have to reconstruct time within a shifting presence" ("Autopoiesis" 11). Such a presentation of "time within time" cannot do otherwise than solder the gap between synchrony and diachrony; it is the very inextricability of the psychic and the social, articulated in meaning, an inextricability whose "identity is difference," that dictates such a temporal suspension of "time within time." For the same reason, the spatial predicament bequeathed by the epistemological concerns of the human sciences—the predicament of observers both outside and inside that which is observed—is characterized by the observation of observation. Formally speaking, observation and the event are versions of each other; just as observation introduces a wedge between the operation "before" (the one being distinguished, rendered visible), and the one to come "thereafter" (that is, the one inaugurated anew with the observing distinction, but which has no present existence for that observation), so the event self-consumes in coming to be, *is* only its unenduring distinctiveness vis-à-vis its past and its future. If the observer appears as his own blindness, that can also be described as the event that appears in its own surpassing. The interpenetration between the spatial (observation) and the temporal (event) descriptions is frankly recognized by Luhmann: "If autopoiesis bases itself on events, a description of the system needs not only one, but two dichotomies: the dichotomy of system and environment and the dichotomy of event and situation" (10).

I reintroduce the term interpenetration in the last sentence advisedly. For it seems to me that here, with the recoding of observational paradoxes as temporal ones, we reach the furthest limit of Luhmann's theoretical endeavor. The system for which such a temporal recoding is most important, even urgent, is not the psychic system, but rather the social system of communication intent on processing information. Jean-François Lyotard has recently offered some intriguing commentary on what he characterizes as the accelerating expansion of a cosmic Leibnizian monad, a *mathesis universalis* governed by the "compulsion to communicate and to secure the communicability of anything at all" (Lyotard 62). For Lyotard, as for Luhmann, the efficiency of the communication system depends to a very large extent on the ability to present time within time, to process and manage the event:

> The importance of the technologies constructed around electronics and information processing resides in the fact that they make the programming and control of memorizing, i.e., the synthesis of different times in one time, less dependent on the conditions of life on earth. It is very probable that among the material complexes we know, the human brain is the most capable of producing complexity in its turn, as the production of the new technologies proves. And as such, it also remains the supreme agency for controlling these technologies.
>
> And yet its survival requires that it be fed by a body, which in turn can only survive in the conditions of life on earth, or a simulacrum of those conditions. I think that one of the essential objectives of research today is to overcome the obstacle that the body places in the way of the development of communicational technologies, i.e., the new extended memory. (62)

When Luhmann characterizes the paradoxes of the social system's autopoiesis as a "built-in requirement of discontinuity and newness" answering the system's *"necessity to handle and process information"* ("Autopoiesis" 10), he describes in his own terms what Lyotard calls the "compulsion to communicate." Social systems as communication processors handle information; they reproduce themselves via the constant rearticulation of the distinction between information and utterance, fact and behavior, themselves and psychic systems. Such a compulsion, we can see, cannot proceed otherwise than by constantly trying to overcome the obstacle of the body, and all the noninformational facets of meaning adhering to psychic and social behavior.

I will return now to Foucault, who described two elements of the analytic of finitude within which we still operate. One was a visual or observational metaphor: "the blind stain" by which it is possible to know man as wedded to what he cannot see. The other was essentially temporal: as the locus of a temporal origin "always in recession to itself." From the perspective of the psychic system, the interpenetration of psychic and social—the expression each gives of the finitude of the other—is most powerfully expressed in the first, observational, metaphor. It is very hard not to place bodies and psyches at the root of all talk of observation. When the temporal paradox comes to the fore—as it does occasionally in Luhmann and Lyotard, for instance—I think we can see the other face of this interpenetration. For it is from the vantage of a truly supra-individual communication system that man's finitude—his very psycho-corporeal embeddedness—is best expressed not as the limit of vision, but as a temporal pulse. More and more, I suspect, we will watch man confront his own finitude not in the form of the mote in his eye, but as the moment that passes in the merest blink. More and more, the human seems less an observer coupled to his own blindness than a switch or relay who can be "on" only because it can just as well be "off." Our metaphors of communication may be giving way to communication's metaphors of us.

Notes

1. See Hacking. Jacques Derrida has also taken up the interrelation between questions of human identity, conjecture, and probability in "Psyche: Inventions of the Other."

2. The list of lectures is given in *Seminar II* (294).

3. The first of these forms would be the "impersonal subject," which is expressed in the "one" of the "one knows that . . ." and which "gives merely the general form of the noetic subject" ("Logical Time" 207). "The second, which is expressed in 'the two whites' who must recognize themselves 'in each other,' introduces the form of the *other as such,* that is as pure reciprocity" (208).

4. See "The Paradoxy of Observing Systems." Speaking of the endless way in which, with each observation, a new unmarked space is at once "severed" and (re)-created, Luhmann comments drily: "We resist the temptation to call this creation."

5. For a clear and up-to-date summary of Luhmann's theory, see Schwanitz.

6. As Habermas has remarked, this theoretical itinerary, precisely in its hostility to the philosophical tradition of the subject of reflection, situates itself as much in that very tradition as anywhere else: "It is not so much the disciplinary tradition of social theory from Comte to Parsons that Luhmann tries to connect up with, as the history of problems associated with the philosophy of the subject from Kant to Husserl. His systems theory does not, say, lead sociology onto the secure path of science; rather it presents itself as the successor to an abandoned philosophy" (Habermas 368).

7. For a subtle consideration of the issue of temporality in Luhmann's work, especially as it relates to deconstructive thought on time, see Cornell.

Works Cited

Cornell, Drucilla. "The Relevance of Time to the Relation between the Philosophy of the Limit and Systems Theory: The Call to Judicial Responsibility." In *The Philosophy of the Limit.* New York: Routledge, 1992.

Derrida, Jacques. "Psyche: Inventions of the Other." Translated by Catherine Porter. In *Reading de Man Reading*, edited by Lindsay Waters and Wlad Godzich. Minneapolis: University of Minnesota Press, 1989.

Forrester, John. "Dead on Time: Lacan's Theory of Temporality." In *The Seductions of Psychoanalysis: Freud, Lacan and Derrida*. Cambridge: Cambridge University Press, 1990.

Foucault, Michel. *The Order of Things: An Archaeology of the Human Sciences*. Translated by Alan Sheridan. New York: Random, 1970.

Habermas, Jürgen. "Excursus on Luhmann's Appropriation of the Philosophy of the Subject through Systems Theory." In *The Philosophical Discourse of Modernity: Twelve Lectures*, translated by Frederick G. Lawrence. Cambridge: MIT Press, 1987.

Hacking, Ian. *The Taming of Chance*. Cambridge: Cambridge University Press 1990.

Heims, Steve Joshua. *Constructing a Social Science for Postwar America: The Cybernetics Group, 1946–1953*. Cambridge: MIT Press, 1991.

Herbert, Christopher. *Culture and Anomie: Ethnographic Imagination in the Nineteenth Century*. Chicago: University of Chicago Press, 1991.

Lacan, Jacques. "De nos antécédents." In *Ecrits*. Paris: Editions du Seuil, 1966.

———. "Le séminaire sur 'La Lettre volée.'" In *Ecrits*. Paris: Editions du Seuil, 1966.

———. "Le Temps logique et l'assertion de certitude anticipée: Un noveau sophisme." In *Ecrits*. Paris: Editions du Seuil, 1966.

———. "Remarque sur le rapport de Daniel Lagache: Psychanalyse et structure de la personnalité." In *Ecrits*. Paris: Editions du Seuil, 1966.

———. *The Seminar of Jacques Lacan*. Book II, *The Ego in Freud's Theory and in the Technique of Psychoanalysis, 1954–55*. Edited by Jacques-Alain Miller and translated by Sylvana Tomaselli. New York: Norton, 1988.

———. "Subversion of the Subject and the Dialectic of Desire in the Freudian Unconscious." *Ecrits: A Selection*. Translated by Alan Sheridan. New York: Norton, 1977.

Lévi-Strauss, Claude. *Introduction to the Work of Marcel Mauss*. Translated by Felicity Baker. London: Routledge and Kegan Paul, 1987.

Luhmann, Niklas. "The Autopoiesis of Social Systems." In *Essays on Self-Reference*. New York: Columbia University Press, 1990.

———. "The Cognitive Program of Constructivism and a Reality That Remains Unknown." In *Selforganization: Portrait of a Scientific Revolution*, edited by Wolfgang Krohn, Günter Küppers, and Helga Nowotny. Dordrecht: Kluwer, 1990.

———. "The Evolutionary Differentiation between Society and Interaction."

In *The Micro-Macro Link,* edited by Jeffrey C. Alexander, Bernhard Giesen, Richard Münch, and Neil J. Smelser. Berkeley: University of California Press, 1987.

———. "How Can the Mind Participate in Communication?" In *Materialities of Communication,* edited by Hans Ulrich Gumbrecht and K. Ludwig Pfeiffer and translated by William Whobrey. Stanford: Stanford University Press, 1994.

———. "The Individuality of the Individual: Historical Meanings and Contemporary Problems." In *Essays on Self-Reference.* New York: Columbia University Press, 1990.

———. "Meaning as Sociology's Basic Concept." In *Essays on Self-Reference.* New York: Columbia University Press, 1990.

———. "The Paradox of Observing Systems." *Cultural Critique* 31 (fall 1995): 37–55.

Lyotard, Jean-François. "Time Today." In *The Inhuman: Reflections on Time,* translated by Geoffrey Bennington and Rachel Bowlby. Stanford: Stanford University Press, 1991.

Schwanitz, Dietrich. "Systems Theory According to Niklas Luhmann—Its Environment and Conceptual Strategies." *Cultural Critique* 30 (spring 1995): 137–70.

Weber, Samuel. *Return to Freud: Jacques Lacan's Dislocation of Psychoanalysis.* Cambridge: Cambridge University Press, 1991.

Wiener, Norbert. *Cybernetics, or Control and Communication in the Animal and the Machine.* 2d ed. Cambridge: MIT Press, 1948.

11. Making Contingency Safe for Liberalism: The Pragmatics of Epistemology in Rorty and Luhmann

Cary Wolfe

What must immediately surprise any reader new to the discourses of systems theory or what is sometimes called "second-order cybernetics" is the rather systematic reliance of this new theoretical paradigm on the figure of vision and, more specifically, observation. That surprise might turn into discomfort if not alarm for readers in the humanities who cut their teeth on the critical genealogy of vision and the Look, which runs, in its modernist incarnation, from Freud's discourse on vision in *Civilization and Its Discontents* through Sartre's *Being and Nothingness* to Lacan's seminars and finally to recent influential work in psychoanalysis and feminist film theory.[1] With the possible exception of Michel Foucault, no recent intellectual has done more to call into question the trope of vision than America's foremost pragmatist philosopher, Richard Rorty. From his groundbreaking early work *Philosophy and the Mirror of Nature* onward, Rorty has argued that the figure of vision in the philosophical and critical tradition is indissoluably linked with representationalism and realism, with the former's assumption that "'making true' and 'representing' are reciprocal relations: the nonlinguistic item which makes *S* true is the one represented by *S*," and with the latter's "idea that inquiry is a matter of finding out the nature of something which lies outside the web of beliefs and desires," in which "the object of inquiry—what lies outside the organism—has a context of its own, a context which is privileged by virtue of being the object's rather than the inquirer's" (*Objectivity, Relativism, and Truth* 4, 96). Instead, Rorty argues, we should reduce this desire for objectivity to a search for "solidarity" and embrace a philosophical holism of the sort found in Dewey, Wittgenstein, and Heidegger, which holds that "words take their meanings from other words rather than by virtue of their representative character" and "transparency to the real" (*Philosophy* 368). Hence, Rorty rejects the representationalist position and its privileged figure, and argues instead that "Our only usable notion of 'objectivity' is 'agreement' rather than mirroring" (*Philosophy* 191).

Rorty's Deweyan reduction of "objectivity" to "solidarity" neatly disposes of all sorts of traditional philosophical problems, most significantly the problem of relativism, which may now be seen as a "pseudo-problem" for pragmatist philosophy because, as Rorty puts it, "the pragmatist does not have a theory of truth, much less a relativistic one. As a partisan of solidarity, his account of the value of cooperative human inquiry has only an ethical base, not an epistemological or metaphysical one" (*Objectivity* 24). On this point, the usefulness of the pragmatist reduction is particularly winning in Rorty's hands. In response to the charge that philosophy in its postmodern and/or pragmatist incarnation automatically becomes relativist once it has ceased to be foundationalist, Rorty responds:

> The view that every tradition is as rational or as moral as every other could be held only by a god, someone who had no need to use (but only to mention) the terms "rational" or "moral," because she had no need to inquire or deliberate. Such a being would have escaped from history and conversation into contemplation and metanarrative. To accuse postmodernism of relativism is to try to put a metanarrative in the postmodernist's mouth. (*Objectivity* 202)

Thus, the Deweyan reduction of objectivity to solidarity provides the ethical basis for the pragmatist's Wittgensteinian epistemology, which insists that "it is contexts all the way down" and that "'grasping the thing itself' is not something that precedes contextualization, but is at best a *focus imaginarius*" (*Objectivity* 100). It would appear, then, that for Rorty the "outside" of belief or description—what used to be called the "referent"—is always already *inside*, insofar as meaning, to borrow Walter Benn Michaels's formulation, is not filtered through what we believe, but is rather *constituted* by what we believe ("Saving the Text" 780).

But the problem with *this* position—a problem that Rorty recognizes—is that it immediately raises the suspicion that "antirepresentationalism is simply transcendental idealism in linguistic disguise ... one more version of the Kantian attempt to derive the object's determinacy and structure from that of the subject" (*Objectivity* 4). Critics of antirepresentationalism imagine "some mighty immaterial force called 'mind' or 'language' or 'social practice' ... which shapes facts out of indeterminate goo," and so "The problem for antirepresentationalists," Rorty continues, is to find a way of putting their point which carries no such suggestion. Antirepresentationalists need to insist that "determinacy" is not what is in question—that . . . it is no truer that "atoms are what they are because we use 'atom' as we do" than that "we use 'atom' as we do because atoms are as they are." (*Objectivity* 5)

But even if we agree with Rorty that "determinacy" is not exactly the issue here, the question of how the outside can be accounted for *at all* certainly is. What *is* the philosophical status, exactly, of those atoms whirling beyond the deterministic ken of our descriptions? Discussing the work of Sellars and Davidson, Rorty writes that "what shows us that life is not just a dream, that our beliefs are in touch with reality, is the *causal*, non-intentional, non-representational links between us and the rest of the universe" (*Objectivity* 159). Pragmatists do indeed accept "the brute, inhuman, causal stubbornness of the gold or the text. But they think this should not be confused with, so to speak, an *intentional* stubbornness, an insistence on being *described in a certain way*, its *own* way" (*Objectivity* 83). Were Rorty's account to end here, we would indeed be left with the outside as black box, as that which must somehow be acknowledged and posited but without letting it do any representational work.

In fact, though, there is a further wrinkle in Rorty's treatment of this problem, and it is crucial to his avoidance of the double bind that plagues some of the more well-known neopragmatist accounts of belief. Rorty imagines the recalcitrant realist responding that the pragmatist, given her account, cannot "find out anything about objects at all," that "You never get outside your own head." Rorty replies, in one of the more disarming moments in the book, that "What I have been saying amounts to accepting this gambit." But, he hastens to add, one of the most central beliefs held by the pragmatist is that "lots of objects she does not control are continually causing her to have new and surprising beliefs." Hence, "She is no more free from pressure from the outside, no more tempted to be 'arbitrary,' than anyone else" (*Objectivity* 101).

In contrast to what we might call the "hard" version of belief propounded by neopragmatists such as Steven Knapp, Walter Benn Michaels, and Stanley Fish—which holds that once you have a belief you will inhabit it "without reservation," with "no distance" (Mitchell 25, 113, 116)—Rorty here provides a "soft" account of belief, one in which beliefs are *always* held "with reservations" because they are held in a world in which (to use William James's picturesque phrase) things are constantly "boiling over" our beliefs about them and impelling us to revise those beliefs (James 106–7). In the parlance of cybernetics and systems theory, which we will explore below, for any given code (or belief), the only source of new information is the as yet uncoded, the random and unpatterned, the "noise" of the outside. Rorty's point is that you may believe whatever you like, but that belief itself—and here is the pragmatist imperative—will have consequences (in this particular instance, mostly *bad* consequences) because it is subject to "pressure from the outside." This is the sense, I think, of Davidson's assertion, which Rorty quotes approvingly, that "most of our beliefs are true"—because we are still around to talk about them! (*Objectivity* 9–10).

From this vantage, the imperative to theory—reflection on belief—derives not from an essentialist "appetite of the mind" (to use William James's phrase)[2] nor a desire for transcendence in either its realist or idealist incarnation (as Knapp and Michaels would have it in their *Against Theory* polemic), but rather from the strategic, adaptive, *pragmatic* value of theory, which any act of intellection will ignore only at its own peril. One might insist, with William James, that the desire to theorize is "characteristically human," but "this would be like saying that the desire to use an opposable thumb remains characteristically human. We have little choice," Rorty continues, "but to use that thumb, and little choice but to employ our ability to recontextualize" (*Objectivity* 110). Thus, the pragmatist "takes off from Darwin rather than from Descartes, from beliefs as adaptations to the environment rather than as quasi-pictures" (*Objectivity* 10); he thinks "of linguistic behavior as tool-using, of language as a way of grabbing hold of causal forces and making them do what we want, altering ourselves and our environment to suit our aspirations" (*Objectivity* 81). In this way, pragmatism "switches attention from 'the demands of the object' to the demands of the purpose which a particular inquiry is supposed to serve." "The effect," Rorty concludes,

> is to modulate philosophical debate from a methodologico-ontological key into an ethico-political key. For now one is debating what purposes are worth bothering to fulfill, which are more worthwhile than others, rather than which purposes the nature of humanity or of reality obliges us to have. For antiessentialists, all possible purposes compete with one another on equal terms, since none are more "essentially human" than others. (*Objectivity* 110)

But here, at precisely this juncture, the radically pluralist imperative of Rorty's pragmatist commitment to contingency begins to break down—or more specifically, begins to be recontained by a more familiar, more complacent and uncritical sort of pluralism. For it may be true, as Rorty puts it, that *"holism takes the curse off naturalism"* (*Objectivity* 109), but no sooner does it resituate the *philosophical* problems of naturalism in an "ethico-political key" than it creates enormous *political* problems by reinscribing Rorty's project within the horizon of a debilitating ethnocentrism and, beyond that, liberal humanism. Rorty's description and defense of ethnocentrism in "Solidarity or Objectivity?" begins by sounding commonsensical enough: "For now to say that we must work by our own lights, that we must be ethnocentric, is merely to say that beliefs suggested by another culture must be tested by trying to weave them together with beliefs we already have" (*Objectivity* 26). But the issue that remains submerged here—and that remained submerged in the lengthy passage we quoted above—is this: Just who *is* this generic "we" in Rorty's discourse? That problem works its way to the surface later in the same essay, where Rorty writes, again in a seemingly commonsensical moment:

> The pragmatists' justification of toleration, free inquiry, and the quest for undistorted communication can only take the form of a comparison between societies which exemplify these habits and those which do not, leading up to the suggestion that nobody who has experienced both would prefer the latter. . . . Such justification is not by reference to a criterion, but by reference to various detailed practical advantages. (*Objectivity* 29)

Even if we leave aside the gesture toward "undistorted communication" (a gesture that Rorty himself has critiqued in Habermas[3]), and even if we ascribe it to the bourgeois liberal values that Rorty inventories, the question that never gets asked is whether all members of Rorty's society experience these "detailed practical advantages" in the way that Rorty imagines. These liberal values and freedoms may extend to all in the abstract—that is, in theory—but do they in *fact*, in *practice*? Clearly the answer is no. This need not lead us to reject out of hand the liberal values Rorty regularly invokes; it is simply to point out that when Rorty claims that "we" should encourage the "end of ideology" (*Objectivity* 184), that "anti-ideological liberalism is, in my view, the most valuable tradition of American intellectual life" (*Objectivity* 64), Rorty is staging a claim that is itself ideological through and through—and this, moreover, in terms easily legible by the light of the most basic sort of Marxist critique of ideology familiar since *The German Ideology*: that Rorty represents the interests and "detailed practical advantages" of his own class, "postmodern bourgeois intellectuals," as the interests of the entire society, rather than acknowledging—as his critics of various stripe have urged him to—that those freedoms and advantages enjoyed by him and by his class are not only *not* shared by those beneath him on the social and economic ladder, but are, it could be argued, in fact purchased at their expense, built upon the very prosperity extracted from their exploitation and alienation and distributed with gross and systematic inequity in the economic and social sphere. What Rorty does not recognize, in other words, is that there is a fundamental contradiction between his putative desire to extend liberal advantages to an ever larger community, and the fact that those advantages are possible for some only because they are purchased at the expense of others. This is simply to say that liberal freedoms in the context of global capitalism rest in large part upon *economic* power and freedom, and that both are operative within a social context of scarcity.

Rorty may bridle at Frank Lentricchia's claim in *Criticism and Social Change* that "our society is mainly unreasonable," a charge to which Rorty responds "unreasonable by comparison to what other society?" (*Objectivity* 15). But what Rorty does not see—and this is the force of Lentricchia's charge—is that Rorty's society, while not perhaps unreasonable for those in Rorty's position, is indeed unreasonable for many millions less fortunate than he. As Nancy Fraser puts it, the problem with "the communitarian comfort of a single 'we'" is that

> Rorty homogenizes social space, assuming tendentiously that there are no deep social cleavages capable of generating conflicting solidarities and opposing 'we's.' It follows from this assumed absence of fundamental social antagonisms that politics is a matter of everyone pulling together to solve a common set of problems. Thus, social engineering can replace political struggle. Disconnected tinkerings with a succession of allegedly discrete social problems can replace transformation of the basic institutional structure. (104)[4]

In this light, it is deeply symptomatic that Rorty relies upon the language of "democracy" and "community," whose homogenizing connotations mask and submerge the "unevenness" in the social and economic sphere that a very different language—the language of "capital" and "class" or, as Luhmann will claim, of functional differentiation—would force to light.

To Rorty's credit, he tries in his most recent work to confront the problems raised by his ethnocentrism. If "we heirs of the Enlightenment think of enemies of liberal democracy like Nietzsche or Loyola as, to use Rawls's word, 'mad'" (*Objectivity* 187), then, he acknowledges, "suddenly we liberal democrats are faced with a dilemma," for "[t]o refuse to argue about what human beings should be like seems to show a contempt for the spirit of accommodation and tolerance, which is essential to democracy." But Rorty quickly dispenses with this dilemma by insisting that "[a]ccommodation and tolerance must stop short of a willingness to work within any vocabulary that one's interlocutor wishes to use" (*Objectivity* 190). Again, Rorty's position here may seem reasonable enough, but the problem is that those who are declared beyond the pale of reason during the course of Rorty's recent work include not only Nietzsche and Loyola, but also Gilles Deleuze, Jean-François Lyotard, Michel Foucault, and all those whom Rorty calls, in a recent *New York Times* op-ed piece,[5] the "unpatriotic left" of the American academy, which "refuses to rejoice in the country it inhabits" and "repudiates the idea of a national identity, and the emotion of national pride"—all those who have experienced "an apparent loss of faith in liberal democracy" (*Objectivity* 220).

Rorty constantly invokes the liberal intellectual's dedication to expanding the range of democratic privileges, freedoms, and values, but what becomes clear in his recent work is that such an expansion can take place only *after* the democratic *ethnos* has been purified of the sort of dissent it needs to encourage. Rorty wants a pluralism that is not *too* plural, a democracy that is not *too* democratic to call into question the values and benefits enjoyed by those of Rorty's class and cultural status. Time and again, rather than enlisting those with very basic, fundamental disagreements into the "conversation" of liberal justice and democracy, he declares them out of the loop—declares them nonsubjects, "fanatics" or "fantastics"—to begin with. And what kind of "conversation" is that? Rorty is quite clear on this point: to "accept the fact that we have to start from where we are," he writes, means that "there are lots of views which we simply cannot take seriously. . . . [W]e can *understand* the revolutionary's suggestion that a sailable boat can't be made out of the planks which make up ours, and that we must simply abandon ship. But we cannot take this suggestion seriously." For Rorty, to take this as a "live option" is not to be a rational partner in the democratic conversation, but—and here the red herring is hard to miss—to be one of those "people who have always hoped to become a New Being, who have hoped to be converted rather than persuaded" (*Objectivity* 29).

Thus, the openness of liberal democratic process is purchased at the expense of the *closedness* of the liberal democratic community, and the "conversation" of liberal justice excludes from the very beginning those with whom substantive differences might be discussed. From this vantage, it is clear, as Cornel West puts it, that Rortyan pragmatism "only

kicks the philosophic props from under liberal bourgeois capitalist societies; it requires no change in our cultural and political practices." Rorty's
project, "though pregnant with possibilities . . . refuses to give birth to the
offspring it conceives. Rorty leads philosophy to the complex world of
politics and culture, but confines his engagement to transformation in
the academy and to apologetics for the modern West" (206–7). In the end,
then, Rorty's philosophical commitment to contingency and the radical
pluralism it promises is tightly recontained by his liberal humanism, and
hence we are forced to say that in Rorty's pragmatism, representationalism is indeed undone on the philosophical level, but only to reemerge in
more powerful and insidious form on the plane of the political.[6]

As we have seen, Rortyan pragmatism moves to front and center the revisable, self-critical, and reflexive nature of all beliefs and descriptions,
but only to recontain that commitment to contingency within an ideology of liberalism that declares out of the picture from the outset (to turn
the ocular metaphor back upon Rorty) those social others whose very
otherness or difference might lead to the critical assessment of one's own
belief. Hence, the Rortyan view gives us no way to theorize the *productive*
and *necessary* relationship between antagonistic beliefs in the social
sphere. It is on the terrain of this last problem that Niklas Luhmann's systems theory will take a decisive step beyond what Cornel West has called
American pragmatism's "evasion of epistemology-centered philosophy,"
and in the process will make it clear that a philosophical commitment
to theorizing the pragmatics of contingency needs *more* epistemology-
centered philosophy, not less.

   For my purposes here, Luhmann's key innovation in this connection is
his theorization of "the observation of observation," which attempts to
make use of the ocular metaphor by divorcing it from the sorts of unifying and representationalist designs that are critiqued by Rortyan philosophy only to reappear in Rortyan politics. For Luhmann, all observations
are constructed atop a constitutive paradox or tautology that those systems
cannot acknowledge and at the same time engage in self-reproduction.
All systems, in other words, are constituted by a necessary "blind spot"
that only *other* observing systems can see and disclose, and the process of
social reproduction depends upon the unfolding, distribution, and circulation of these constitutive paradoxes and tautologies (which would
otherwise block systemic self-reproduction) by a plurality of observing
systems. Both Luhmann and Rorty begin from the Wittgensteinian position that "a system," as Luhmann puts it, "can see only what it can see. It

cannot see what it cannot" (*Ecological Communication* 23). But Luhmann, unlike Rorty, derives from this formulation not the irrelevance of other observing systems (or beliefs)—not their exclusion from the conversation of social reproduction—but rather their very necessity.

Luhmann's theorization of the concept of observation and its relation to contingency is heavily indebted to the pioneering work in biology and epistemology of Humberto Maturana and Francisco Varela, and specifically to their theorization of what they term "autopoiesis." The cornerstone distinction for the theory of autopoiesis is between "organization" and "structure," or what Luhmann will call "system" and "element," both of which roughly (but only roughly) correspond to Rorty's closed circuit of "belief" on the one hand and the open, "causal" and "nonintentional" set of relations between belief and the world of "facts" on the other. As Maturana and Varela explain it, "*Organization* denotes those relations that must exist among the components of a system for it to be a member of a specific class"; *structure*, on the other hand, "denotes the components and relations that actually constitute a particular unity and make its organization real" (*Tree* 46–47). For Maturana and Varela, what characterizes all living things is that the relationship between structure and organization is one of "*autopoietic organization*," that is, "they are continually self-producing" (*Tree* 43) according to their own internal rules and requirements. In more general terms, what this means is that all autopoietic entities are *closed*— or, to employ Luhmann's preferred term, "self-referential"—on the level of *organization*, but *open* on the level of *structure*.

This is most clear, perhaps, in Maturana and Varela's theorization of what they call "*operational closure*." "It is interesting to note," they write,

> that the operational closure of the nervous system tells us that it does not operate according to either of the two extremes: it is neither representational nor solipsistic.
>
> It is not solipsistic, because as part of the nervous system's organism, it participates in the interactions of the nervous system with its environment. These interactions continuously trigger in it the structural changes that modulate its dynamics of states. . . .
>
> Nor is it representational, for in each interaction it is the nervous system's structural state that specifies what perturbations are possible and what changes trigger them. (*Tree* 169)

The theorization of the operational closure of autopoietic entities allows Maturana and Varela to break with the last vestiges of the representationalist view and to "walk on the razor's edge, eschewing the extremes of representationalism (objectivism) and solipsism (idealism)" (*Tree* 241). "These two extremes," Varela et al. contend in *The Embodied Mind*, "both take representation as their central notion: in the first case representation is used to recover what is outer; in the second case it is used to project what is inner" (172). As Maturana and Varela frame the problem in *The Tree of Knowledge*, "If we deny the objectivity of a knowable world," they ask, "are we not in the chaos of total arbitrariness because everything is possible?" The way "to cut this apparent Gordian knot," as they put it, is to change the nature of the question, to realize that the first principle of any sort of knowledge whatsoever is that "everything said is said by some-one" (135)—that is, to foreground the contingency of what Luhmann will call *observation*.

Luhmann's refinement of the concept of observation is a key compo-nent of his extension of the concept of autopoiesis from the realm of liv-ing systems (the focus of Maturana and Varela) to social systems as well. "If we abstract from life and define autopoiesis as a general form of sys-tem building using self-referential closure," Luhmann writes, "we would have to admit that there are nonliving autopoietic systems" (*Essays on Self-Reference* 2). For Luhmann as for Maturana and Varela, the attrac-tion of the concept of autopoeisis—or what Luhmann will more often treat under the term "self-reference"—is not least of all that the theoriza-tion of systems as both (operationally) closed and (structurally) open accounts for both high degrees of systemic autonomy *and* the problem of how systems change and "adapt" to their environments (or achieve "reso-nance" with them, as Luhmann puts it in *Ecological Communication*).

But Luhmann extends and refines the work of Maturana and Varela in the particular theoretical pressure he applies to the problem of observa-tion. It will come as no surprise that Luhmann agrees with Maturana and Varela that "[a]utopoietic systems . . . are sovereign with respect to the constitution of identities and differences. They, of course, do not create a material world of their own. They presuppose other levels of reality. . . . But whatever they use as identities and as differences is of their own mak-ing" (*Essays* 3). But in his essay "Complexity and Meaning," Luhmann pushes beyond Maturana and Varela and distinguishes between a system's *operation* and its *observation*. "By operation," he writes, "I mean the actual processing of the reproduction of the system." "By observation, on the other hand," he continues, "I mean the act of distinguishing for the crea-tion of information" (*Essays* 83). The distinction between operation and observation, Luhmann writes elsewhere, "occupies the place that had been taken up to this point by the unity-seeking logic of reflection. (*This means, therefore, a substitution of difference for unity)*"—about which we will say much more in a moment ("Cognitive Program" 68, emphasis added).

Luhmann distinguishes a third term here as well: *self-observation.* "Self-referential systems are able to observe themselves," he writes. "By using a fundamental distinction schema to delineate their self-identities, they can direct their own operations toward their self-identities" (*Essays* 123). If they do not do so—if they cannot distinguish what is systemic and internal from what is environmental and external—then they cease to exist as autopoietic, self-producing systems. This is why Luhmann writes that the distinction between "internal" and "external" observation "is not needed," that "the concept of observation includes 'self-observation'" (*Essays* 82). In other words, to observe *at all* requires an autopoietic system, and an autopoietic system capable of observation cannot exist without the capacity for self-observation—that is, without the capacity "to handle distinctions and process information."[7] Hence, observation and, within that, self-observation, are *themselves* necessary operations of autopoietic systems.

All of which leads us to the central point we need to understand about Luhmann's concept of observation and its relationship to the epistemological problem of constructivism. Luhmann's position is clearest, perhaps, in his explanation of the observation of observation in his important essay "The Cognitive Program of Constructivism and a Reality That Remains Unknown," where he writes, "An operation that uses distinctions in order to designate something we will call 'observation.' We are caught once again, therefore, in a circle: the distinction between operation and observation appears itself as an element of observation" (68–69). Most readers would probably agree with Luhmann—and beyond that, with the work of George Spencer Brown, which Luhmann draws upon—that the most elementary intellectual and psychical act is to draw a distinction, to distinguish figure from ground, "x" from "not-x." The point Luhmann wishes to underscore, however, has been a familiar one ever since the "liar's paradox" of antiquity (*Ecological Communication* xiv), or more recently Russell and Whitehead's theory of "logical types," which tried to solve such antinomies[8]: that drawing such a distinction, which is the elementary constitutive act of observation, is always either paradoxical or tautological, and that this is both necessary and unavoidable. "Tautologies are distinctions," Luhmann writes,

> that do not distinguish. They explicitly negate that what they distinguish really makes a difference. Tautologies thus block observations. They are

always based on a dual observation schema: something is what it is. This statement, however, negates the posited duality and asserts an identity. Tautologies thus negate what makes them possible in the first place, and, therefore, the negation itself becomes meaningless. (*Essays* 136)

To many readers, this description will evoke nothing so much as the famous Hegelian postulate of "the identity of identity and non-identity." What Luhmann wishes to stress, however, is not the identity of identity and nonidentity, but rather the *nonidentity* (or difference) of identity and nonidentity. As he puts it in *Ecological Communication,*

> the *unity* (of self-reference) that would be unacceptable in the form of a tautology (e.g., legal is legal) or a paradox (one does not have the legal right to maintain their legal right) is replaced by a *difference* (e.g., the difference of legal and illegal). Then the system can proceed according to this difference, oscillate within it and develop programs to regulate the ascription of the operations of the code's positions and counterpositions without raising the question of the code's unity. (xiv)

Two points need to be stressed here. First, what enables this crucial emphasis on the *difference* of identity and nonidentity—it is also what separates Luhmann from the Kantianism with which he bears more than a passing resemblance—is Luhmann's strident rejection of any possibility of a transcendental subject-observer. For Luhmann, all observations are produced by a *contingent* observer who could always, in theory, describe things otherwise. Hence, all observations—and all systems described by them—contain an irreducible element of complexity. As William Rasch puts it, for Luhmann—*contra* Hegel and Kant—"complexity can never be fully reduced to an underlying simplicity since simplicity, like complexity, is a construct of observation that could always be other than it is. Contingency, the ability to alter perspectives, acts as a reservoir of complexity within all simplicity" (70).[9]

The second point that needs to be underscored in reference to Luhmann's position on tautology that we quoted above—and it is one whose pragmatic impulse will distinguish Luhmann's position from that of Derrida and deconstruction, at least in Luhmann's eyes[10]—is the insistence that the tautological (or, more strictly, paradoxical[11]) nature of all observation constitutes a real, pragmatic *problem* for all social self-descriptions. This is so, Luhmann argues, because "[a]n observer can realize the self-referential systems are constituted in a paradoxical way. This insight itself, however, makes observation impossible, since it postulates an autopoietic system whose autopoiesis is blocked" (*Essays* 139).

The solution to these obstacles or blockages is that self-referential paradoxes must be—in Luhmann's somewhat frustrating nomenclature—"unfolded" by the system. We have already mentioned two ways in which such unfolding might take place: the theory of logical types, which "interrupts" or unfolds the vicious circle of paradoxical self-reference "by an arbitrary fiat: the instruction to ignore operations that disobey the command to avoid paradoxes" (*Ecological* 24); and the reliance upon binary coding, which enables the system to deparadoxize itself by orienting the operations of the system to the *difference* of x and not-x (legal and nonlegal, for example) without raising the question of their paradoxical identity.

But if Luhmann's concern with the pragmatics of tautology and paradox for social reproduction separates him from Derrida and deconstruction, his position on how the practical-political "unfolding" of tautology and paradox ought to be handled separates him from consensus-seeking liberals such as Rorty or Habermas. For if the processes of "deparadoxization" require that a system's constitutive paradox remain invisible to it, then the only way that this fact can be known as such is by the observation of *another* observing system. As Luhmann puts it, "Only an [other] observer is able to realize what systems themselves are unable to realize" (*Essays* 127). What is decisive about Luhmann's intervention here is his insistence on the constitutive blindness of all observations, a blindness that does not separate or alienate us from the world but, paradoxically, guarantees our connection with it.

As Luhmann explains it, in a remarkable passage worth quoting at length:

> The source of a distinction's guaranteeing reality lies in its own operative unity [as, for example, legal versus not legal]. It is, however, precisely as this unity that the distinction cannot be observed—except by means of another distinction which then assumes the function of a guarantor of reality. Another way of expressing this is to say the operation emerges simultaneously with the world which as a result remains cognitively unapproachable to the operation.
>
> The conclusion to be drawn from this is that the connection with the reality of the external world is established by the blind spot of the cognitive operation. *Reality is what one does not perceive when one perceives it.* ("Cognitive Program" 76, emphasis added)

Perception and cognition of reality, in other words, are made possible by the deployment of a paradoxical distinction to which the observation utilizing that distinction must remain "blind" to perceive and cognize *at all.* Here, Luhmann neatly traverses what has traditionally seemed an insoluble epistemological problem: how to avoid the untenable reliance upon the science/ideology distinction that has traditionally buttressed ideology critique and the sociology of knowledge, and at the same time avoid lapsing into epistemological solipsism. Luhmann's negotiation of this problem is possible *only* on the strength of systems theory's articulation of the observation of observation, which enables us to view the "blind spot" or "latency" of the observations of others not merely as ideological bias or the distortion of a pregiven reality knowable by "science," but rather as the unavoidably partial and paradoxical precondition of knowing as such.[12] This, Luhmann writes, is "the systematic keystone of epistemology—taking the place of its *a priori* foundation" ("Cognitive Program" 75). "In a somewhat Wittgensteinian formulation," he writes,

> one could say that a system can see only what it can see. It cannot see what it cannot. Moreover, it cannot see that it cannot see this. . . .
>
> Nevertheless, a system that observes other systems has other possibilities. . . . [T]he observation of a system by another system—following Humberto Maturana we will call this "second-order observation"—can also observe the restrictions forced on the observed system by its own mode of operation. . . . It can observe the horizons of the observed system so that what they exclude becomes evident. (*Ecological Communication* 23)

And here, we need to sharpen our sense of the pragmatic implications of Luhmann's epistemology and how it differs from Rortyan pragmatism. The passage we quoted a moment ago—that "the constructed reality is not . . . the reality referred to"—must surely remind us of Rorty's attempt to situate descriptions within a "nonintentional" and "causal" world *without* having that world do representational work. But what follows—that "the connection with the reality of the external world is established by the blind spot of the cognitive operation," that "[r]eality is what one does not perceive when one perceives it"—separates Luhmann's crucial reformulation from Rorty. For Luhmann stresses the contingency and paradoxicality of that very observation itself and—*contra* Rorty— derives from that contingency the necessity of the observations of others: It is only in the mutual observations of *different* observers that a critical view of any observed system can be formulated. If we are stuck with distinctions that are paradoxical and must live with blind spots at the heart of our observations, Luhmann writes, "Perhaps, then, the problem can be distributed among a plurality of interlinked observers" who are of necessity joined to the world and to each other by their constitutive but different blind spots. The work of social theory would then consist in developing "thoughtful procedures for observing observation, with a special emphasis on that which, for the other, is a paradox and, therefore, cannot be observed by him" ("Sthenography" 137).

And while this reformulation is neither, strictly speaking, a politics nor an ethics, it *does* provide a rigorous and persuasive theorization of the compelling necessity of *sociality as such*—that is, of necessary reciprocal and yet asymmetrical relations between self and other, observer and observed, relations that can no longer be characterized in terms of an identity principle (be it that of class, race, or what have you) that would reduce the full complexity and contingency of the observer's position in the social space.

In these terms, Luhmann's insistence on the "blind spot" of observation and, therefore, the essential aporia of any authority that derives from it (the authority, say, of the system that enforces the distinction legal/illegal) bears more than a passing resemblance to the proposition of a fundamental "antagonism" at the core of social relations as recently theorized by Ernesto Laclau, Chantal Mouffe, and Slavoj Zizek. As Zizek articulates the concept, "far from reducing all reality to a kind of language-game, the socio-symbolic field is conceived as structured around a certain traumatic impossibility, around a certain fissure which *cannot* be symbolized" ("Beyond Discourse-Analysis" 249). Or, to remind ourselves of Luhmann's formulation, "the connection with the reality of the external world is established by the blind spot of the cognitive operation. Reality is what one does not perceive when one perceives it." For Zizek, as for Luhmann, "every identity is already in itself blocked, marked by an impossibility" (252), and thus "the stake of the entire process of subjectivation, of assuming different subject-positions"—or in Luhmann's system, of a plurality of interlinked observers whereby paradox and tautology can be distributed in the social field—"is ultimately to enable us to avoid this traumatic experience" (253) of the fact, as Luhmann puts it, that it is our blind spot that assures our connection with the real, that "[t]he constructed reality is . . . not the reality referred to."[13]

For Zizek, the concept of social antagonism, which countenances "an ethics of confrontation with an impossible, traumatic kernel not covered by any ideal," constitutes "the only real answer to Habermas, to the project based on the ethics of the ideal of communication without constraint," because it unmasks the constitutive disavowal at work in Habermas's model: "I know very well that communication is broken and perverted, but still . . . (I believe and act as if the ideal speech situation is already realized)" (259). For Habermas, we will remember, complexity and contingency always contain the threat of relativism and even nihilism, and thus the proliferation of different systems of knowledge and value must be grounded in some sort of underlying simplicity. For Habermas—but not, significantly, for Rorty[14]—that simplicity is harbored in the very nature of language itself and its fundamental presupposition of an ideal speech act, of undistorted communication through which the claims of different systems of thought and value can be adjudicated in a process of rational dialogue that arrives at common norms and values (Rasch 70–72). But Zizek, like Luhmann, does *not* disavow the "broken and perverted" (i.e., paradoxical and tautological) nature of communication, but rather derives from that brokenness the necessity of sociality as such. He holds that "what this fetishistic logic of the ideal is masking, is of course, the limitation proper to the symbolic field as such: the fact that the signifying field is always structured around a certain fundamental deadlock" (259) or what Luhmann calls the "blockage" of paradoxical self-reference.

Like the theorists of social antagonism, then—and like them, against Habermas *and* against Rortyan ethnocentrism—Luhmann insists that the distribution of the problem of paradoxicality and the circulation of latent possibilities can take place only if we do *not* opt for the quintessentially modernist and Enlightenment strategy of the hoped-for *reduction* of complexity via social consensus. If all observation is made possible by a paradoxical distinction to which it must remain blind, then

> This is why all projection, or the setting of a goal, every formation of episodes necessitates recursive observation and why, furthermore, recursive observation makes possible not so much the elimination of paradoxes as their temporal and social distribution onto different operations. A consensual integration of systems of communication is, given such conditions, something that should sooner be feared than sought for. For such integration can only result in the paradoxes becoming invisible to all and remaining that way for an indefinite future. ("Cognitive Program" 75)

For Luhmann, the Habermasian strategy—or, for that matter, the Rortyan one of liberal recontainment of contingency via ethnocentrism—is a doomed and potentially dangerous project that might result in the blockage of communications and the "invisibilizing," rather than the unfolding and distribution, of paradox. Clearly, then, the Luhmannian concept of observation is not "intended to provide a grounding for knowledge, but only to keep open the possibility of observation operations being carried out by very different empirical systems—living systems, systems of consciousness, systems of communications" ("Cognitive" 78). And just as clear, too, is Luhmann's resolute posthumanism, which concludes that what Habermas characterizes as the project of Enlightenment and modernity has—and must—come to an end. "With this," he writes,

> the traditional attribution of cognition to "man" has been done away with. It is clear here, if anywhere, that "constructivism" is a completely new theory of knowledge, a post-humanistic one. This is not intended maliciously but only to make clear that the concept "man" (in the singular!), as a designation for the bearer and guarantor of the unity of knowledge, must be renounced. The reality of cognition is to be found in the current operations of the various autopoietic systems. ("Cognitive" 78)

There is a pragmatic premium in this philosophical difference, for in Luhmann's view the movement to a posthumanist perspective has the practical benefit of enabling "better functional performance" (*Ecological* 128) of highly differentiated society and its component systems. For example, in *Ecological Communication,* Luhmann argues that "a sensible handling of system-theoretical analyses" will "lead more to the expansion of the perspectives of problems than to their suppression" (131). Such analysis, he contends, can provide an important counterbalance to destructive social anxiety, which "is more likely to stop the effects of society on its environment, but . . . has to pay for this by risking unforeseeable internal reactions that again produce anxiety" (131)—and here, we might think of the so-called spotted owl controversy, where social anxiety about biodiversity and habitat destruction did indeed "stop the effects of society on its environment," but at the expense of creating a severe generalized backlash of anxiety about environmental protection at the expense of economic well-being, one that threatened, ironically enough, to have severe repercussions for the reauthorization by Congress in 1993–94 of the Endangered Species Act, the very act that mandated the protection of the spotted owl in the first place!

It is important to note, however, that Luhmann makes it abundantly clear in many, many places that the pragmatic value of his theorization of complexity and functional differentiation is to enable *this world*—and more specifically this liberal Western capitalist world—to engage in systemic self-reproduction without destructive blockages of autopoiesis, the better to achieve maximum resonance between the system and its environment. Luhmann—and this is quite surprising, given his epistemological innovation—wholly takes for granted the enclosure of thought, even putatively revolutionary thought, by the Western liberal capitalist social system. As he puts it in *Political Theory in the Welfare State,* the basic problem for any would-be critical position is that

> every operational act, every structural process, every partial system participates in the society, and is society, but in none of these instances is it possible to discern the existence of the whole society. Even the criticisms of society must be carried out within society. Even the planning of society must be carried out within society. Even the description of society must be carried out within society. (17)

And while Luhmann would seem to register here nothing more than an epistemological truism, in fact he goes a good bit farther—as Danilo Zolo has pointed out—in his tacit endorsement of liberal capitalist society and "neo-liberal" policies (a fact more than hinted at in Luhmann's political essays and in his systematically reductive glances at Marxist theory).[15] As Zolo puts it, Luhmann interprets

> the crisis of the welfare state in terms of the loss of the law's regulating ability. Accordingly, legislation invades private spheres as well as other functionally differentiated and autonomous sub-systems. In doing so, the welfare state's interventionist strategy overloads the law to the point of distorting its regulatory function. This overload results in chaotic legislation which complicates the legal system and prevents its rational self-reproduction. Against this, Luhmann and the reflexive law theorists defend the autopoietic autonomy of social sub-systems—particularly those concerning economy, education, and family life. Thus, the autopoietic paradigm supports deregulatory policies. (63)

To recall our discussion of Rorty, then, we may say that Luhmann, while he *does* evade pragmatism's "evasion of philosophy" and its reduction of complexity, he *does not* evade a pervasive liberalism which, even more than in Rorty, takes the form of a technocratic functionalism that is content to operate wholly within the purview of what Lyotard has called the "performativity principle" of "positivist pragmatism" (66).[16] In these terms, John McGowan's recent critique of Rorty would surely apply to Luhmann as well. As McGowan puts it, "the important thing to note is that the negative endorsement of change, of the ever continuing conversation"—or, we should add, of the continual unfolding of complexity and distribution of paradox in Luhmann's system—"is dependent upon and presupposes a much more positive version of the social world that the conversationalists inhabit" (198). And it is here that Luhmann's complacent taking for granted of Western capitalist liberal society short-circuits the second political promise of his work: his rigorous theorization of the epistemological necessity of sociality as such, of the fact that the social is always virtual, partial, and perspectival, mutually constituted by observers who can and must expose the aporias of each other's positions.

This shortcoming will be clearest, perhaps, if we compare with Luhmann Donna Haraway's reinterpretation of the figure of "observation" and, more broadly, of vision in her recent work. Both Luhmann and Haraway attempt to retheorize the figure of vision by *situating* it—that is, by de-transcendentalizing it and divorcing it from its representationalist associations. Luhmann would, I think, agree with Haraway's insistence on "the embodied nature of all vision" and her theoretical rejection of "a conquering gaze from nowhere," one which claims "the power to see and not be seen, to represent while escaping representation" (188). And like Luhmann, Haraway's epistemological project is dedicated above all— to use her paraphrase of Althusser—to resisting "simplication in the last instance" (196). But here, Haraway's sense of "embodiment" as the name for this theoretical fact needs to be distinguished from Luhmann's theorization of the contingency of all observation. What Haraway wants is a concept of "situated knowledges" (188), which emphasizes the physical and social positionality of the observer—not least of all, for Haraway, the observer's gender—the specific conjuncture of qualities that mark the possibilities and limits of what the observer can see. In this sense, she writes, "objectivity turns out to be about particular and specific embodiment, and definitely not about the false vision promising transcendence of all limits and responsibility. The moral is simple: only partial perspective promises objective vision" (190). In Haraway's articulation of observation and vision, "embodiment" names contingency, "objectivity" names political and ethical responsibility for one's observations, and both are "as hostile to various forms of relativism as to the most explicitly totalizing versions of claims to scientific authority" (191).

There can be little doubt that Haraway would find in Luhmann's theorization of observation—his "unmarking" of it, we might say, through relentless formalism and abstraction that socially and historically disembody it—confirmation of her suspicions about relativism. Luhmann would need to be told, as Haraway reminds us, that "social constructivism cannot be allowed to decay into the radiant emanations of cynicism" (184). And indeed, Luhmann would seem to invite this charge— both theoretically and tonally, rhetorically—in many places in his work. In *Ecological Communication*, for example, he writes:

The problem seems to be that one has to recognize the dominant social structure—whether seen as "capitalism" or "functional differentiation"—to assume a position against it. . . . A functional equivalent for the [nineteenth-century] theoretical construct "dialectics/revolution" is not in sight and therefore it is not clear what function a critical self-observation of society within society could fulfill. . . . Like the "Reds" . . . the "Greens" will also lose color as soon as they assume office and find themselves confronted with all the red tape. (126)

My guess is that Haraway would detect—and would be justified in detecting—the leveling political extrapolation at the end of this passage from the epistemological claims at its beginning as an instance of that relativism which is, in her words,

a way of being nowhere and everywhere equally. The "equality" of positioning is a denial of responsibility and critical enquiry. Relativism is the perfect mirror twin of totalization in the ideologies of objectivity; both deny the stakes in location, embodiment, and partial perspective. (191)

Luhmann's theory of observation doesn't sufficiently recognize the imperative of Haraway's "embodied objectivity": that "vision is *always* a question of the power to see" (192). Again, a passage from Luhmann's *Ecological Communication* will help to make the point:

Investigations that are inspired theoretically can always be accused of a lack of "practical reference." They do not provide prescriptions for others to use. . . . This does not exclude the possibility that serviceable results can be attained in this way. But then the significance of theory will always remain that a more controlled method of creating ideas can increase the probability of more serviceable results—above all, that it can reduce the probability of creating useless excitement. (xviii)

The question that Haraway puts squarely on the table is never broached by Luhmann: "serviceable" *for whom*? And in the absence of addressing that question—and of any detectable *interest* in addressing it—Luhmann's position seems ripe for interpellation into Haraway's reading of systems theory in terms of the historically specific "management" strategies of post–World War II liberal capitalist society, in which systems theory, like sociobiology, population genetics, ergonomics, and other field models, is crucial to "the reproduction of capitalist social relations" in the specific era of "an engineering science of automated technological devices, in which the model of scientific intervention is technical and 'systematic,' [t]he nature of analysis is technological functionalism, and ideological appeals are to alleviation of stress and other signs of human obsolescence" (44).

We need not agree with Haraway, I think, that the systems theory paradigm always already carries with it a tacit endorsement of liberal capitalist society in its post–World War II incarnation. Indeed, Maturana and Varela have drawn very different ethical and political implications from very similar epistemological premises.[17] But it seems clear that in Luhmann's hands, the systems theory paradigm does in fact indulge the same sort of blithe liberal functionalism embraced by Rorty in its refusal to confront the uneven and asymmetrical relations of power—especially economic power—which undeniably constrain and indeed often render utterly beside the point the unfolding of complexity and the distribution of paradox that remain in Luhmann's thought too squarely within a political if not philosophical idealism. If Rorty sanitizes the social field by limiting conversation to the liberal *ethnos,* Luhmann levels it by refusing to complicate his epistemological pluralism—that we are all alike in the formal homology of our observational differences—with an account of how in the real social world in which those observations take place, some observers enjoy more resources of observation than others. The complexifying and open-ended imperative of Luhmann's theory is, following George Spencer Brown, "distinguish!" and "observe!" but we must still subject that imperative to the critique leveled by Steven Best and Douglas Kellner at the metaphor of cultural "conversation" of diversity and plurality as it is deployed by Rorty: "that some people and groups are in far better positions—politically, economically, and psychologically—to speak" (or to observe, we might add) "than others. Such calls are vapid," they continue, "when the field of discourse is controlled and monopolized by the dominant economic and political powers" (288).

We might say, then, that Luhmann's "blind spot"—his unobservable constitutive distinction—is his unspoken distinction between "differentiation" and what historicist, materialist critique has theorized as "contradiction," a blind spot that manifests itself in Luhmann's inability or unwillingness to adequately theorize the discrepancy between the formal equivalence of observers in his epistemology and their real lack of equivalence on the material, economic plane. It seems that the category of contradiction—insofar as it names precisely this difference—proves more difficult to dispose of than Luhmann's systems theory imagines. Or rather—to put a somewhat finer point on it—it is disposed of by systems theory, but only "abstractly," as Marxist theorists like to say, only in thought, but not in historical, material practice. What Luhmann's epistemological idealism refuses to confront is that the differentiation, autonomy, and unfolding of complexity it imagines remain muffled and mastered by the economic context of identity and exchange value within which systems theory itself historically arises. And in that refusal, in its pragmatic effect of socially reproducing the liberal status quo, it is clear that there are powerful ideological reasons, as well as epistemological ones, why one cannot see what one cannot see.

Notes

1. See, for example, Slavoj Zizek's post-Lacanian analysis of the Look in a few different texts, most importantly *The Sublime Object of Ideology* (London: New Left Books, 1989), and, within feminism, the wealth of work by critics such as Mary Anne Doane, Laura Mulvey, Kaja Silverman, and many others.

2. See Lentricchia's discussion of this moment in James—as in pointed contrast to the *Against Theory* polemic—in his *Ariel and the Police* (124–33).

3. See Rorty, "Habermas and Lyotard on Postmodernity."

4. This is not to agree with Fraser's Habermasian call for a renewed attention to the normative. But Fraser's point is strikingly borne out in Rorty's response to Clifford Geertz's critique of his ethnocentrism, where Rorty declares in response to Geertz's example that the "whole apparatus of the liberal democratic state" functions just fine when it ensures that the drunken Indian needing dialysis in Geertz's example is "going to have more years in which to drink than he would otherwise have had," that the point for doctors and lawyers in such cases is— and here the quintessence of the sort of technocratic functionalism that Fraser describes—"to get their job done, and to do it right" (*Objectivity* 204–5).

5. "The Unpatriotic Academy," *New York Times*, 13 February 1994, sec. E, p. 15.

6. In this light, Rortyan pluralism seems subject to the description of Gilles Deleuze's critique of "state philosophy" offered by Brian Massumi: "More insidious than its institution-based propogation is the State-form's ability to propagate *itself* without centrally directed inculcation (liberalism and good citizenship). Still more insidious is the process presiding over our present plight, in which the moral and philosophical foundations of national and personal identity have crumbled, making a mockery of the State-form—but the world keeps right on going *as if* they hadn't." See Brian Massumi, *A User's Guide*, p. 5.

7. Luhmann qualifies this somewhat in the essay "Complexity and Meaning": "[I]t has to be decided," he writes, "whether self-observation (or the capacity to handle distinctions and process information) is a prerequisite of autopoietic systems" (*Essays* 82). It seems, though, that the position outlined earlier in the essay—

that the concept of observation automatically includes that of self-observation—would seem to require that self-observation is such a prerequisite.

8. Luhmann addresses the so-called "theory of logical types" of Russell and Whitehead in many places; see, for example, the essay "Tautology and Paradox" (*Essays* 127) and, for a more extensive refutation, *Ecological Communication* (23–24).

9. As Rasch points out, this is precisely the point that is missed by Habermas's project of a universal pragmatics. "The whole movement of Habermas's thought," he writes, "tends to some final resting place, prescriptively in the form of consensus as the legitimate basis for social order, and methodologically in the form of a normative underlying simple structure which is said to dictate the proper shape of surface complexity" (78).

10. In a recent essay, Luhmann writes of the lineage that runs from Nietzsche through Heidegger to Derrida that in their work "[p]aradoxicality is not avoided or evaded but, rather, openly exhibited and devotedly celebrated. . . . At present, it is not easy to form a judgment of this. Initially, one is impressed by the radicality with which the traditional European modes of thought are discarded. . . . [But it] has so far not produced significant results. The paradoxicalization of civilization has not led to the civilization of paradoxicality. One also starts to wonder whether it is appropriate to describe today's extremely dynamic society in terms of a semantic that amounts to a mixture of arbitrariness and paralysis" ("Sthenography" 134).

11. As Luhmann points out, tautologies are actually "special cases of paradoxes"; "tautologies turn out to be paradoxes, while the reverse is not true" (*Essays* 136).

12. As he puts it in "Cognitive Program," "The assumption—to be found above all in the classical sociology of knowledge—that latent structures, functions and interests lead to distortions of knowledge, if not to blatant errors, [and] can and must be abandoned. The impossibility of distinguishing the distinction that one distinguishes with is an unavoidable precondition of cognition. The question of whether a given choice of distinction suits one's latent interests only arises on the level of second-order observation [that is, on the level of the observation of observation]" (73).

13. This is to leave aside, of course, the pronounced differences between Zizek and Luhmann: Zizek's conjugation of these issues in a psychoanalytic register in which the concepts of trauma and affect are crucial; and the fact that Zizek's project remains very much within the terms of Cartesian or Kantian idealism, even if it inverts those terms so that the symbolic, signifier, or idea now appears as the failed "gentrification" of the Real, "the thing," the body—of all that Kant in the *Critique of Practical Reason* calls "the pathological."

14. For a sketch of their differences, see Rorty's essay "Habermas and Lyotard on Postmodernity." As Rorty puts it, "the trouble with Habermas is not so much that he provides a metanarrative of emancipation as that he feels the need to legitimize, that he is not content to let the narratives which hold our culture together do their stuff. He is scratching where it does not itch" (164).

15. See, for example, Luhmann's discussion of the relationship between politics and economics in *Political Theory in the Welfare State,* pp. 11–19; for a typically reductive glance at Marxist theory, see pp. 17–18.

16. See also pp. 123–24. For a particularly striking instance of Rorty's technocratic functionalism, see his response to Clifford Geertz's critique of his ethnocentrism in *Objectivity,* pp. 203–10.

17. For a discussion of the ethical and political implications of their epistemology, see the last chapter of *The Tree of Knowledge,* especially pp. 245–46 and, in this volume, Cary Wolfe, "In Search of Posthumanist Theory: The Second-Order Cybernetics of Maturana and Varela."

Works Cited

Best, Steven, and Douglas Kellner. *Postmodern Theory: Critical Interrogations.* New York: Guilford Press, 1991.

Fraser, Nancy. *Unruly Practices: Power, Discourse and Gender in Contemporary Social Theory.* Minneapolis: University of Minnesota Press, 1989.

Haraway, Donna J. *Simians, Cyborgs, and Women: The Reinvention of Nature.* New York: Routledge, 1991.

James, William. *Pragmatism and the Meaning of Truth.* Introduction by A. J. Ayer. Cambridge: Harvard University Press, 1978.

Lentricchia, Frank. *Ariel and the Police: William James, Michel Foucault, Wallace Stevens.* Madison: University of Wisconsin Press, 1988.

———. *Criticism and Social Change.* Chicago: University of Chicago Press, 1983.

Luhmann, Niklas. "The Cognitive Program of Constructivism and a Reality That Remains Unknown." In *Selforganization: Portrait of a Scientific Revolution,* edited by Wolfgang Krohn, Gunter Kuppers, and Helga Nowotny. Dordrecht: Kluwer Academic Publishers, 1990.

———. *Ecological Communication.* Translated by John Bednarz. Chicago: University of Chicago Press, 1989.

———. *Essays on Self-Reference.* New York: Columbia University Press, 1990.

———. *Political Theory in the Welfare State.* Translated by John Bednarz Jr. Berlin: Walter de Gruyter, 1990.

———. "Sthenography." Translated by Bernd Widdig. *Stanford Literature Review* 7 (1990): 132–36.

Lyotard, Jean-François. *The Postmodern Explained: Correspondence 1982–1985.* Edited by Julian Pefanis and Morgan Thomas and translated by Don Barry et al. Afterword by Wlad Godzich. Minneapolis: University of Minnesota Press, 1992.

Massumi, Brian. *A User's Guide to Capitalism and Schizophrenia: Deviations from Deleuze and Guattari.* Cambridge: MIT Press, 1992.

Maturana, Humberto R., and Francisco J. Varela. *The Tree of Knowledge: The Biological Roots of Human Understanding.* Rev. ed. Translated by Robert Paolucci. Foreword by J. Z. Young. Boston: Shambhala Press, 1992.

McGowan, John. *Postmodernism and Its Critics.* Ithaca: Cornell University Press, 1991.

Michaels, Walter Benn. "Saving the Text: Reference and Belief." *MLN* 93 (1978): 765–89.

Mitchell, W. J. T., ed. *Against Theory: Literary Studies and the New Pragmatism.* Chicago: University of Chicago Press, 1985.

Rasch, William. "Theories of Complexity, Complexities of Theory: Habermas, Luhmann, and the Study of Social Systems." *German Studies Review* 14 (1991): 56–79.

Rorty, Richard. "Habermas and Lyotard on Postmodernity." In *Habermas and Modernity,* edited by Richard Bernstein. Cambridge: MIT Press, 1985.

———. *Objectivity, Relativism, and Truth.* Cambridge: Cambridge University Press, 1991.

———. *Philosophy and the Mirror of Nature.* Princeton, N.J.: Princeton University Press, 1979.

Varela, Francisco J., Evan Thompson, and Eleanor Rosch. *The Embodied Mind: Cognitive Science and Human Experience.* Cambridge: MIT Press, 1993.

West, Cornel. *The American Evasion of Philosophy: A Genealogy of Pragmatism.* Minneapolis: University of Minnesota Press, 1989.

Zizek, Slavoj. "Beyond Discourse-Analysis." In *New Reflections of the Revolution of Our Time,* by Ernesto Laclau. London: New Left Books, 1990.

———. *The Sublime Object of Ideology.* London: Verso, 1989.

Zolo, Danilo. "Autopoiesis: Critique of a Postmodern Paradigm." *Telos* 86 (winter 1990–91): 60–84.

12. The Autonomy of Affect

Brian Massumi

I

> A man builds a snowman on his roof garden. It starts to melt in the after-
> noon sun. He watches. After a time, he takes the snowman to the cool of
> the mountains, where it stops melting. He bids it good-bye, and leaves.

Just images, no words, very simple. It was a story depicted in a short film
shown on German TV as a fill-in between programs. The film drew complaints from parents reporting that their children had been frightened.
That drew the attention of a team of researchers. Their study was notable
for failing to find much of what it was studying: cognition.

Researchers, headed by Hertha Sturm, used three versions of the film:
the original wordless version and two versions with voice-overs added.
The first voice-over version was dubbed "factual." It added a simple step-
by-step account of the action as it happened. A second version was called
"emotional." It was largely the same as the "factual" version, but included
at crucial turning points words expressing the emotional tenor of the
scene under way.

Sets of nine-year-old children were tested for recall, and asked to rate
the version they saw on a scale of "pleasantness." The factual version was
consistently rated the least pleasant, and was also the worst remembered.
The most pleasant was the original wordless version, which was rated just
slightly above the emotional. And it was the emotional version that was
best remembered.

This is already a bit muddling. Something stranger happened when the
subjects of the study were asked to rate the individual scenes in the film
simultaneously on a "happy-sad" scale and a "pleasant-unpleasant" scale.
The "sad" scenes were rated the *most pleasant,* the sadder the better.

The hypothesis that immediately suggests itself is that in some kind of precocious anti-Freudian protest, the children were equating arousal with pleasure. But this being an empirical study, the children were wired. Their physiological reactions were monitored. The factual version elicited the highest level of arousal, even though it was the most unpleasant (i.e., happy) and made the least long-lasting impression. The children, it turns out, were physiologically split: factuality made their heart beat faster and deepened their breathing, but it made their skin resistance fall. The original nonverbal version elicited the greatest response from their skin. Galvanic skin response measures *autonomic* reaction.

From the tone of their report, it seems that the researchers were a bit taken aback by their results. They contented themselves with observing that the difference between sadness and happiness is not all that it's cracked up to be, and worrying that the difference between children and adults was also not all that it was cracked up to be (judging by studies of adult retention of news broadcasts). Their only positive conclusion was *the primacy of the affective* in image reception (Sturm 25–37).

Accepting and expanding upon that, it could be noted that the primacy of the affective is marked by a gap between *content* and *effect*: it would appear that the strength or duration of an image's effect is not logically connected to the content in any straightforward way. This is not to say that there is no connection and no logic. What is meant here by the content of the image is its indexing to conventional meanings in an intersubjective context, its sociolinguistic qualification. This indexing fixes the *quality* of the image; the strength or duration of the image's effect could be called its *intensity*. What comes out here is that there is no correspondence or conformity between quality and intensity. If there is a relation, it is of another nature.

To translate this negative observation into a positive one: the event of image reception is multileveled, or at least bilevel. There is an immediate bifurcation in response into two seemingly autonomous systems. One, the level of intensity, is characterized by a crossing of semantic wires: on it, sadness is pleasant. The level of intensity is organized according to a logic that does not admit of the excluded middle. This is to say that it is not semantically or semiotically ordered; it does not fix distinctions. Instead, it vaguely but insistently connects what is normally indexed as separate. When asked to signify itself, it can only do so in a paradox. There is disconnection of signifying order from intensity—which constitutes a different order of connection operating in parallel. The gap noted earlier is not only between content and effect; it is also between the form of content—signification as a conventional system of distinctive difference—and intensity. The disconnection between form/content and intensity/effect is not just negative: it enables a different connectivity, a different difference, in parallel.

Both levels, qualification and intensity, are immediately embodied. Intensity is embodied in purely autonomic reactions most directly manifested in the skin—at the surface of the body, at its interface with things. Depth reactions belong more to the form/content (qualification) level, even though they also involve autonomic functions such as heartbeat and breathing. The reason may be that they are associated with expectation, which depends on consciously positioning oneself in a line of narrative continuity. Modulations of heartbeat and breathing mark a reflux of consciousness into the autonomic depths, coterminous with a rise of the autonomic into consciousness. They are a conscious-autonomic mix, a measure of their participation in one another. Intensity is beside that loop, a nonconscious, never-to-conscious autonomic remainder. It is outside expectation and adaptation, as disconnected from meaningful sequencing, from narration, as it is from vital function. It is narratively delocalized, spreading over the generalized body surface, like a lateral backwash from the function-meaning interloops traveling the vertical path between head and heart.

Language, though headstrong, is not simply in opposition to intensity. It would seem to function differentially in relation to it. The factual version of the snowman story was dampening. Matter-of-factness dampens intensity. In this case, matter-of-factness was a doubling of the sequence of images with a narration expressing in as objective a manner as possible the commonsense function and consensual meaning of the movements perceived on screen. This interfered with the images' effect. The emotional version added a few phrases that punctuated the narrative line with qualifications of the emotional content, as opposed to the objective-narrative content. The qualifications of emotional content enhanced the images' effect, as if they resonated with the level of intensity rather than interfering with it. An emotional qualification breaks narrative continuity for a moment to register a state—actually re-register an already felt state (for the skin is faster than the word).

The relationship between the levels of intensity and qualification is not one of conformity or correspondence, but of resonation or interference, amplification or dampening. Linguistic expression can resonate with and amplify intensity at the price of making itself functionally redundant. When, on the other hand, it doubles a sequence of movements in order to add something to it in the way of meaningful progression—in this case a sense of futurity, expectation, an intimation of what comes next in a conventional progression—then it runs counter to and dampens the intensity. Intensity would seem to be associated with nonlinear processes: resonance and feedback that momentarily suspend the linear progress of the narrative present from past to future. Intensity is qualifiable as an emotional state, and that state is static—temporal and narrative noise. It is a state of suspense, potentially of disruption. It's like a temporal sink, a hole in time, as we conceive of it and narrativize it. It is not exactly passivity, because it is filled with motion, vibratory motion, resonation. And it is not yet activity, because the motion is not of the kind that can be directed (if only symbolically) toward practical ends in a world of constituted objects and aims (if only on screen). Of course the qualification of an emotion is quite often, in other contexts, itself a narrative element that moves the action ahead, taking its place in socially recognized lines of action and reaction. But to the extent that it is, it is not in resonance with intensity. It resonates to the exact degree to which it is in excess of any narrative or functional line.

In any case, language doubles the flow of images, on another level, on a different track. There is a redundancy of resonation that plays up or amplifies (feeds back disconnection, enabling a different connectivity), and a redundancy of signification that plays out or linearizes (jumps the feedback loop between vital function and meaning into lines of socially valorized action and reaction). Language belongs to entirely different orders depending on which redundancy it enacts. Or, it always enacts both more or less completely: two languages, two dimensions of every expression, one superlinear, the other linear. Every event takes place on both levels—and between both levels, as they themselves resonate to form a larger system composed of two interacting subsystems following entirely different rules of formation. For clarity, it might be best to give different names to the two halves of the event. In this case: *suspense* could be distinguished from and interlinked with *expectation,* as superlinear and linear dimensions of the same image-event, which is at the same time an expression-event.

Approaches to the image in its relation to language are incomplete if they operate only on the semantic or semiotic level, however that level is defined (linguistically, logically, narratologically, ideologically, or all of these in combination, as a Symbolic). What they lose, precisely, is the expression *event*—in favor of structure. Much could be gained by integrating the dimension of intensity into cultural theory. The stakes are the new. For structure is the place where nothing ever happens, that explanatory heaven in which all eventual permutations are prefigured in a self-consistent set of invariant generative rules. Nothing is prefigured in the event. It is the collapse of structured distinction into intensity, of rules into paradox. It is the suspension of the invariance that makes happy happy, sad sad, function function, and meaning mean. Could it be that it is through the expectant suspension of that suspense that the new emerges? As if an echo of irreducible excess, of gratuitous amplification, piggybacked on the reconnection to progression, bringing a tinge of the unexpected, the lateral, the unmotivated, to lines of action and reaction. A change in the rules. The expression-event is the system of the inexplicable; emergence, into and against (re)generation (the re-production of a structure). In the case of the snowman, the unexpected and inexplicable that emerged along with the generated responses had to do with the differences between happiness and sadness, children and adults, not being all they're cracked up to be, much to our scientific chagrin: a change in the rules. Intensity is the unassimilable.

For present purposes, intensity will be equated with affect. There seems to be a growing feeling within media, literary, and art theory that affect is central to an understanding of our information and image-based late-capitalist culture, in which so-called master narratives are perceived to have foundered. Fredric Jameson notwithstanding, belief has waned for many, but not affect. If anything, our condition is characterized by a surfeit of it. The problem is that there is no cultural-theoretical vocabulary specific to affect.[1] Our entire vocabulary has derived from theories of signification that are still wedded to structure even across irreconcilable differences (the divorce proceedings of poststructuralism: terminable or interminable?). In the absence of an asignifying philosophy of affect, it is all too easy for received psychological categories to slip back in, undoing the considerable deconstructive work that has been effectively carried out by poststructuralism. Affect is most often used loosely as a synonym for emotion.[2] But one of the clearest lessons of this first story is that emotion and affect—if affect is intensity—follow different logics and pertain to different orders.

An emotion is a subjective content, the sociolinguistic fixing of the quality of an experience which is from that point onward defined as personal. Emotion is qualified intensity, the conventional, consensual point of insertion of intensity into semantically and semiotically formed progressions, into narrativizable action-reaction circuits, into function and meaning. It is intensity owned and recognized. It is crucial to theorize the difference between affect and emotion. If some have the impression that it has waned, it is because affect is unqualified. As such, it is not ownable or recognizable, and is thus resistant to critique.

It is not that there is no philosophical antecedents to draw on. It is just that they are not the usual ones for literary and cultural studies. On many of these points there is a formidable philosophical precursor: on the difference in nature between affect and emotion; on the irreducibly bodily and autonomic nature of affect; on affect as a suspension of action-reaction circuits and linear temporality in a sink of what might be called "passion," to distinguish it both from passivity and activity; on the equation between affect and effect; on the form/content of conventional discourse as constituting an autonomous or semiautonomous stratum running counter to the full registering of affect and its affirmation, its positive development, its expression as and for itself—on all of these points, it is the name of Baruch Spinoza that stands out. Much is to be gained by a rereading of Spinoza. And the title of his central work suggests a designation for the project of thinking affect: Ethics.[3]

II

Another story, about the brain: the mystery of the missing half second.

Experiments were performed on patients who had been implanted with cortical electrodes for medical purposes. Mild electrical pulses were administered to the electrode and also to points on the skin. In either case, the stimulation was felt only if it lasted more than half a second: half a second, the minimum perceivable lapse. If the cortical electrode was fired a half second before the skin was stimulated, patients reported feeling the skin pulse first. The researcher speculated that sensation involves a "backward referral in time"—in other words, that sensation is organized recursively before being linearized, before it is redirected outward to take its part in a conscious chain of actions and reactions. Brain and skin form a resonating vessel. Stimulation turns inward, is folded into the body, except that there is no inside for it to be in, because the body is radically open, absorbing impulses quicker than they can be perceived, and because the entire vibratory event is unconscious, out of mind. Its anomaly is smoothed over retrospectively to fit conscious requirements of continuity and linear causality.[4]

What happens during the missing half second? A second experiment gave some hints.

Brain waves of healthy volunteers were monitored by an electroencephalograph (EEG) machine. The subjects were asked to flex a finger at a moment of their choosing, and to note the time of their decision on a clock. The flexes came 0.2 seconds after they clocked the decision. But the EEG machine registered significant brain activity 0.3 seconds *before* the decision. Again, a half second lapse between the beginning of a bodily event and its completion in an outwardly directed, active expression.

Asked to speculate on what implications all this might have for a doctrine of free will, the researcher Benjamin Libet "proposes that *we may exert free will not by initiating intentions but by vetoing, acceding or otherwise responding to them after they arise*" (Horgan).

In other words, the half second is missed not because it is empty, but because it is overfull, in excess of the actually performed action and of its ascribed meaning. Will and consciousness are *subtractive*. They are *limitative, derived functions* that reduce a complexity too rich to be functionally expressed. It should be noted in particular that during the mysterious half second, what we think of as "higher" functions, such as volition, are apparently being performed by autonomic, bodily reactions occuring in the brain but outside consciousness, and between brain and finger, but prior to action and expression. The formation of a volition is necessarily accompanied and aided by cognitive functions. Perhaps the snowman researchers of the first story couldn't find cognition because they were looking for it in the wrong place—in the "mind," rather than in *the body* they were monitoring. Talk of intensity inevitably raises the objection that such a notion inevitably involves an appeal to a prereflexive, romantically raw domain of primitive experiential richness—the nature in our culture. It is not that. First, because something happening out of mind in a body directly absorbing its outside cannot exactly be said to be experienced. Second, because volition, cognition, and presumably other "higher" functions usually presumed to be in the mind, figured as a mysterious container of mental entities that is somehow separate from body and brain, are present and active in that now not-so "raw" domain. Resonation assumes feedback. "Higher functions" belonging to the realm of qualified form/content in which identified, self-expressive persons interact in conventionalized action-reaction circuits following a linear time line are fed back into the realm of intensity and recursive causality. The body doesn't just absorb pulses or discrete stimulations; it infolds *contexts,* it infolds volitions and cognitions that are nothing if not situated. Intensity is asocial, but not presocial—it *includes* social elements, but mixes them with elements belonging to other levels of functioning, and combines them according to different logic. How could this be so? Only if the *trace* of past actions *including a trace of their contexts* were conserved in the brain and in the flesh, but out of mind and out of body understood as qualifiable interiorities, active and passive respectively, directive spirit and dumb matter. Only if past actions and contexts were conserved and repeated, autonomically reactivated, but not accomplished; begun, but not completed. Intensity is *incipience,* incipient action and expression. Intensity is not only incipience, but the incipience of mutually exclusive pathways of action and expression that are then reduced, inhibited, prevented from actualizing themselves completely—all but one. Since the crowd of pretenders to actualization are tending toward completion in a new context, their incipience cannot just be a conservation and reactivation. They are *tendencies*—in other words, pastnesses opening onto a future, but with no present to speak of. For the present is lost with the missing half second, passing too quickly to be perceived, too quickly, actually, to have happened.

This requires a complete reworking of how we think about the body. Something that happens too quickly to have happened, actually, is *virtual.* The body is as immediately virtual as it is actual. The virtual, the pressing crowd of incipiencies and tendencies, is a realm of *potential.* In potential is where futurity combines, unmediated, with pastness, where outsides are infolded, and sadness is happy (happy because the press to action and expression is life). The virtual is a lived paradox where what are normally opposites coexist, coalesce, and connect; where what cannot be experienced cannot but be felt—albeit reduced and contained. For out of the pressing crowd an individual action or expression will emerge and be registered consciously. One "wills" it to emerge, to be qualified, to take on sociolinguistic meaning, to enter linear action-reaction circuits, to become a content of one's life—by dint of inhibition.

Since the virtual is unlivable even as it happens, it can be thought of as a form of superlinear abstraction that does not obey the law of the excluded middle, that is organized differently but is inseparable from the concrete activity and expressivity of the body. The body is as immediately abstract as it is concrete; its activity and expressivity extend, as on their underside, into an incorporeal, yet perfectly real, dimension of pressing potential.

Here, too, there is a philosophical precursor: on the brain as a center of indetermination; on consciousness as subtractive and inhibitive; on perception as working to infold extended actions and expressions, *and* their situatedness, into a dimension of intensity or *in*tension as opposed to extension; on the continual doubling of the actual body by this dimension of intensity, understood as a superlinear, superabstract realm of potential; on that realm of the virtual as having a different temporal structure, in which past and future brush shoulders with no mediating present, and as having a different, recursive causality; on the virtual as cresting in a liminal realm of emergence, where half-actualized actions and expressions arise like waves on a sea to which most no sooner return—on all of these points, the name of Henri Bergson imposes itself (see in particular *Matter and Memory*).

Bergson could profitably be read together with Spinoza. One of Spinoza's basic definitions of affect is an "affection (in other words an impingement upon) the body, *and at the same time the idea of the affection.*" This starts sounding suspiciously Bergsonian if it is noted that the body, when impinged upon, is described by Spinoza as being in a state of passional suspension in which it exists more outside of itself, more in the abstracted action of the impinging thing and the abstracted context of that action, than within itself; and if it is noted that the idea in question is not only not conscious, but is not in the first instance in the "mind."

In Spinoza, it is only when the idea of the affection is doubled by an *idea of the idea of the affection* that it attains the level of conscious reflection. Conscious reflection is a doubling over of the idea on itself, a self-recursion of the idea that enwraps the affection or impingement, at two removes. For it has already been removed once, by the body itself. The body infolds the *effect* of the impingement—it conserves the impingement minus the impinging thing, the impingement abstracted from the actual action that caused it and the actual context of that action. This is a first-order idea produced spontaneously by the body: the affection is immediately, spontaneously doubled by the repeatable trace of an encounter, the "form" of an encounter, in Spinoza's terminology (an infolding, or contraction, of context in the vocabulary of this essay). The trace determines a tendency, the potential, if not yet the appetite, for the autonomic repetition and variation of the impingement. Conscious reflection is the doubling over of this dynamic abstraction on itself. The order of connection of such dynamic abstractions among themselves, on a level specific to them, is called mind. The autonomic tendency received second hand from the body is raised to a higher power to become an activity of the mind. Mind and body are seen as two levels recapitulating the same image/expression event in different but parallel ways, ascending by degrees from the concrete to the incorporeal, holding to the same absent center of a now spectral—and potentialized—encounter. Spinoza's Ethics is the philosophy of the becoming-active, in parallel, of mind and body, from an origin in passion, in impingement, in so pure and productive a receptivity that it can only be conceived as a third state, an excluded middle, prior to the distinction between activity and passivity: affect. This "origin" is never left behind, but doubles one like a shadow that is always almost perceived, and cannot but be perceived, in effect.

In a different but complementary direction, when Spinoza defines mind and body as different orders of connection, or different regimes of motion and rest, his thinking converges in suggestive ways with Bergson's theories of virtuality and movement.

When the names Bergson and Spinoza are mentioned together in the same sentence, the name of Gilles Deleuze inevitably follows. It is Deleuze who reopened the path to these authors, although nowhere does he patch them directly into each other. Spinoza, Bergson, and Deleuze could profitably be read together with recent theories of complexity and chaos. It is all a question of *emergence*, which is precisely the focus of the various science-derived theories that converge around the notion of self-organization (the spontaneous production of a level of reality having its own rules of formation and order of connection). Affect or intensity in the present account is akin to what is called a critical point, or a bifurcation point, or singular point, in chaos theory and the theory of dissipative structures. This is the turning point at which a physical system paradoxically embodies multiple and normally mutually exclusive potentials, only one of which is "selected." "Phase space" could be seen as a diagrammatic rendering of the dimension of the virtual. The organization of multiple levels that have different logics and temporal organizations but are locked in resonance with each other and recapitulate the same event in divergent ways recalls the fractal ontology and nonlinear causality underlying theories of complexity.

The levels at play could be multiplied to infinity: already mentioned are mind and body, but also volition and cognition, at least two orders of language, expectation and suspense, body depth and epidermis, past and future, action and reaction, happiness and sadness, quiescence and arousal, passivity and activity. These could be seen not as binary oppositions or contradictions, but as resonating levels. Affect is their point of emergence, in their actual specificity; and it is their vanishing point, in singularity, in their virtual coexistence and interconnection—that critical point shadowing every image/expression-event. Theories of self-organization could help realize Deleuze's project of a "transcendental empiricism." Félix Guattari's last book, *Chaosmose,* explores the intersection between his work, solo and with Deleuze, and chaos theory. The term "transcendental empiricism," however, was dropped early on in Deleuze's works and not taken up again explicitly by Guattari. Although the realm of intensity that Deleuze's philosophy strives to conceptualize is transcendental in the sense that it is not directly accessible to experience, it is not fair to say that it is outside experience either. It is immanent to it—always in it but not of it. Intensity and experience accompany one another, like two mutually presupposing dimensions, or like two sides of a coin. Intensity is immanent to matter and to events, to mind and to body and to every level of bifurcation composing them and which they compose. Thus it also cannot but be experienced, in effect—in the proliferations of levels of organization it ceaselessly gives rise to, generates and regenerates, at every suspended moment. Deleuze's philosophy is the point at which transcendental philosophy flips over into a radical immanentism, and empiricism into ethical experimentation. The Kantian imperative to understand the conditions of possible experience as if from outside and above transposes into an invitation to recapitulate, to repeat and complexify, ground level, the real conditions of emergence, not of the categorical, but of the unclassifiable, the unassimilable, the never yet felt, the felt for less than half a second, again for the first time—the new. Kant meets Spinoza, where idealism and empiricism turn pragmatic, becoming a midwifery of invention—with no loss in abstractive or inductive power. Quite the contrary—both are heightened. But now abstraction is synonymous with an unleashing of potential, rather than its subtraction. And the sense of induction has changed, to a triggering of a process of complexifying self-organization. The implied ethics of the project is the value attached—without foundation, with desire only—to the multiplication of powers of existence, to ever-divergent regimes of action and expression.

Feedback (Digression)

The work of Gilbert Simondon is an invaluable resource for this kind of project.[5] An example is his treatment of the feedback of atoms of "higher" modes of organization into a level of emergence. He sees this functioning even on the physical level, where "germs" of forms are present in an emergent dimension along with unformed elements such as tropisms (attractors), distributions of potential energy (gradients defining metastabilities), and nonlocalized relations (resonance). According to Simondon, the dimension of the emergent—which he terms the "pre-individual"—cannot be understood in terms of form, even if it infolds forms in a germinal state. It can only be analyzed as a continuous but highly differentiated *field* that is "out of phase" with formed entities (has a different topology and causal order from the "individuals" that arise from it and whose forms return to it). A germinal or "implicit" form cannot be understood as a shape or structure. It is more a bundle of potential functions localized, as a differentiated region, within a larger field of potential. The regions are separated from each other by dynamic thresholds rather than by boundaries. Simondon calls these regions of potential "quanta," even as they appear on the macrophysical level, and even on the human level (99) (hence the atomic allusion). Extrapolating a bit, the "regions" are obviously abstract, in the sense that they do not define boundaried spaces, but are rather differentiations within an open field characterized by action at a distance between elements (attractors, gradients, resonation). The limits of the region, and of the entire field (the universe), are defined by the reach of its elements' collective actions at a distance. The limit will not be a sharp demarcation, more like a multidimensional fading to infinity. The field is open in the sense that it has no interiority or exteriority: it is limited *and* infinite.

"Implicit" form is a bundling of potential functions, an infolding or contraction of potential interactions (intension). The playing out of those potentials requires an *unfolding* in three-dimensional space and linear time: extension as actualization, actualization as *expression*. It is in expression that the fade-out occurs. *The limits of the field of emergence are in its actual expression.* Implicit form may be thought of as the effective presence of the sum total of a thing's interactions, minus the thing. It is a thing's relationality autonomized as a dimension of the real. This *autonomization of relation* is the condition under which "higher" functions feed back. Emergence, once again, is a two-sided coin: one side in the virtual (the autonomy of relation), the other in the actual (functional limitation). What is being termed affect in this essay is precisely this two-sidedness, the simultaneous participation of the virtual in the actual and the actual in the virtual, as one arises from and returns to the other. Affect is this two-sideness *as seen from the side of the actual thing*, as couched in its perceptions and cognitions. Affect is *the virtual as point of view*, provided the visual metaphor is used guardedly. For affect is synaesthetic, implying a participation of the senses in each other: the measure of a living thing's potential interactions is its ability to transform the effects of one sensory mode into those of another (tactility and vision being the most obvious but by no means only examples; interoceptive senses, especially proprioception, are crucial).[6] Affects are *virtual synaesthetic perspectives* anchored in (functionally limited by) the actually existing, particular things that embody them. The *autonomy* of affect is its participation in the virtual. *Its autonomy is its openness.* Affect is autonomous to the degree to which it escapes confinement in the particular body whose vitality, or potential for interaction, it is. Formed, qualified, situated perceptions and cognitions fulfilling functions of actual connection or blockage are the *capture* and closure of affect. Emotion is the intensest (most contracted) expression of that capture—and of the fact that something has always and again escaped. Something remains unactualized, inseparable from but unassimilable to any *particular,* functionally anchored perspective. That is why all emotion is more or less disorienting, and why it is classically described as being outside of oneself, at the very point at which one is most intimately and unshareably in contact with oneself and one's vitality. If there were no escape, no excess or remainder, no fade-out to infinity, the universe would be without potential, pure entropy, death. Actually existing, structured things live in and through that which escapes them. Their autonomy is the autonomy of affect.

The escape of affect *cannot but be perceived, alongside* the perceptions that are its capture. This side-perception may be punctual, localized in an event (such as the sudden realization that happiness and sadness are something besides what they are). When it is punctual, it is usually described in negative terms, typically as a form of *shock* (the sudden interruption of functions of actual connection).[7] But it is also continuous, like a background perception that accompanies every event, however quotidian. When the continuity of affective escape is put into words, it tends to take on positive connotations. For it is nothing less than *the perception of one's own vitality,* one's sense of aliveness, of changeability (often signified as "freedom"). One's "sense of aliveness" is a continuous, nonconscious *self-perception* (unconscious self-reflection). It is the perception of this self-perception, its naming and making conscious, that allows affect to be effectively analyzed—as long as a vocabulary can be found for that which is imperceptible but whose escape from perception cannot but be perceived, as long as one is alive.[8]

Simondon notes the connection between self-reflection and affect. He even extends the capacity for self-reflection to all living things (149)—although it is hard to see why his own analysis does not constrain him to extend it to all *things* (is not resonation a kind of self-reflection?). Spinoza could be read as doing this in his definition of the idea of the affection as a trace—one that is not without reverberations. More radically, he sees ideas as attaining their most adequate (most self-organized) expression not in us but in the "mind" of God. But then he defines God as Nature (understood as encompassing the human, the artificial, and the invented). Deleuze is willing to take the step of dispensing with God. One of the things that distinguishes his philosophy most sharply from that of his contemporaries is the notion that ideality is a dimension of matter (also understood as encompassing the human, the artificial, and the invented) (see in particular *Difference and Repetition*).

The distinction between the living and the nonliving, the biological and the physical, is not the presence or absence of reflection, but its directness. Our brains and nervous systems effect the autonomization of relation, in an interval smaller than the smallest perceivable, even though the operation arises from perception and returns to it. In the more primitive organisms, this autonomization is accomplished by organism-wide networks of interoceptive and exteroceptive sense receptors whose impulses are not centralized in a brain. One could say that a jellyfish is its brain. In all living things, the autonomization of relation is effected by a center of indetermination (a localized or organism-wide function of resonation that delinearizes causality in order to relinearize it with a change of direction: from reception to reaction). At the fundamental physical level, there is no such mediation.[9] The place of physical non-mediation between the virtual and the actual is explored by quantum mechanics. Just as "higher" functions are fed back—all the way to the subatomic (i.e., position and momentum)—quantum indeterminacy is fed forward. It rises through the fractal bifurcations leading to and between each of the superposed levels of reality. On each level, it appears in a unique mode adequate to that level. On the level of the physical macrosystems analyzed by Simondon, its mode is potential energy and the margin of "play" it introduces into deterministic systems (epitomized by the three-body problem so dear to chaos theory). On the biological level, it is the margin of undecidability accompanying every perception, which is one with a perception's transmissibility from one sense to another. On the human level, it is that same undecidability fed forward into thought, as evidenced in the deconstructability of every structure of ideas (as expressed, for example, in Gödel's incompleteness theorem and in Derrida's *différance*). Each individual and collective human level has its peculiar "quantum" mode (various forms of undecidability in logical and signifying systems are joined by emotion on the psychological level, resistance on the political level, the specter of crisis haunting capitalist economies, and so forth). These modes are fed back and fed forward into one another, echoes of each other one and all.

The use of concept of the quantum outside quantum mechanics, even as applied to human psychology, is not a metaphor. For each level, it is necessary to find an operative concept for the objective indeterminacy that echoes what on the subatomic level goes by the name of quantum. This involves analyzing every formation as participating in what David Bohm calls an *implicate order* cutting across all levels and doubled on each (Bohm and Hiley). Affect is as good a general term as any for the interface between implicate and explicate order.[10] Turning to the difference between the physical and the biological, it is clear that there can be no firm dividing line between them, nor between them and the human. Affect, like thought or reflection, could be extended to any or every level, providing that the uniqueness of its functioning on that level is taken into account. The difference between the dead, the living, and the human is not a question of form or structure, nor of the properties possessed by the embodiments of forms or structures, nor of the qualified functions performed by those embodiments (their utility or ability to do work). The distinction between kinds of things and levels of reality is a question of degree: of the way in which modes of organization (such as reflection) are differentially present on every level, bar the extremes. The extremes are the quantum physical and the human inasmuch as it aspires to or confuses itself with the divine (which occurs wherever notions of changelessness, eternity, identity, and essence are operative). Neither extreme can be said to exist, although each could be said to be real, in entirely different ways (the quantum is productive of effective reality, and the divine is effectively produced, as a fiction). In between lies a continuum of existence differentiated into levels, or regions of potential, between which there are no boundaries, only dynamic thresholds.

As Simondon notes, all of this makes it difficult to speak of either transcendence or immanence (156). No matter what one does, they tend to flip over into each other, in a kind of spontaneous Deleuzian combustion. It makes little difference if the field of existence (being plus potential; the actual in its relation with the virtual) is thought of as an infinite interiority or a parallelism of mutual exteriorities. You get burned either way. Spinoza had it both ways (an indivisible substance divided into parallel attributes). To the extent that the terms transcendence and immanence connote spatial relations—and they inevitably do—they are inadequate to the task. A philosophical sleight of hand like Spinoza's is always necessary. The trick is to get comfortable with productive paradox.

All of this—the absence of a clear line of demarcation between the physical, the vital, the human, and the superhuman; the undecidability of immanence and transcendence—also has important implications for ethical thought. A common thread running through the varieties of social

constructivism currently dominant in cultural theory holds that everything, including nature, is constructed in discourse. The classical definition of the human as the rational animal returns in a new permutation:
the human as the chattering animal. Only the animal is bracketed: the
human as the chattering of culture. This reinstates a rigid divide between
the human and the nonhuman, since it has become a commonplace,
after Lacan, to make language the special preserve of the human (chattering chimps notwithstanding). Now saying that the quantum level is
transformed by our perception is not the same as saying that it is only *in*
our perception; saying that nature is discursively constructed is not necessarily the same as saying that nature is *in* discourse. Social constructivism easily leads to a cultural solipsism analogous to subjectivist interpretations of quantum mechanics. In this worst case solipsist scenario,
nature appears as immanent to culture (as its construct). At best, when
the status of nature is deemed unworthy of attention, it is simply shunted
aside. In that case it appears, by default, as transcendent to culture (as its
inert and meaningless remainder). Perhaps the difference between best
and worst is not all that it is cracked up to be. For in either case, nature as
naturing, nature as having its own dynamism, is erased. Theoretical
moves aimed at ending Man end up making human culture the measure
and meaning of all things, in a kind of unfettered anthropomorphism
precluding—to take one example—articulations of cultural theory and
ecology. It is meaningless to interrogate the relation of the human to the
nonhuman if the nonhuman is only a construct of human culture, or
inertness. The concepts of nature and culture need serious reworking, in
a way that expresses the irreducible *alterity* of the nonhuman in and
through its active *connection* to the human, and vice versa. It is time that
cultural theorists let matter be matter, brains be brains, jellyfish be jellyfish, and culture be nature, in irreducible alterity and infinite connection.

A final note: the feedback of "higher" functions can take such forms as
the deployment of narrative in essays about the breakdown of narrative.

III

Next story.

The last story was of the brain. This one is of the brainless. His name is
Ronald Reagan. The story comes from a well-known book of pop neuro-
physiology by Oliver Sacks (76–80).

Sacks describes watching a televised speech by the "Great Communicator" in a hospital ward of patients suffering from two kinds of cognitive
dysfunction. Some were suffering from global aphasia, which rendered

them incapable of understanding words as such. They could nonetheless understand most of what was said because they compensated by developing extraordinary abilities to read extraverbal cues: inflection, facial expression, and other gestures—body language. Others on the ward were suffering from what is called tonal agnosia, which is the inverse of aphasia. The ability to hear the expressiveness of the voice is lost, and with it goes attention to other extraverbal cues. Language is reduced to its grammatical form and semantic or logical content. Neither group appeared to be Reagan voters. In fact, the speech was universally greeted by howls of laughter and expressions of outrage. The "Great Communicator" was failing to persuade. To the aphasiacs, he was functionally illiterate in extraverbal cuing; his body language struck them as hilariously inept. He was, after all, a recycled bad actor, and an aging one at that. The agnosiacs were outraged that the man couldn't put together a grammatical sentence or follow a logical line to its conclusion. He came across to them as intellectually impaired. (It must be recalled that this is long before the onset of Reagan's recently announced Alzheimer's disease—what does that say about the difference between normality and degeneration?)

Now all of this might have come as news to those who think of Reagan and other postmodern political stars on the model of charismatic leadership, in which the fluency of a public figure's gestural and tonal repertoire mesmerize the masses, lulling them into bleary-eyed belief in the content of the mellifluous words. On the contrary, what is astonishing is that Reagan wasn't laughed and jeered off the campaign podium and was swept into office not once but twice. It wasn't that people didn't hear his verbal fumbling or recognize the incoherence of his thoughts. They were the butt of constant jokes and news stories. And it wasn't that what he lacked on the level of verbal coherence was glossed over by the seductive fluency of his body image. Reagan was more famous for his polyps than his poise, and there was a collective fascination with his faltering health and regular shedding of bits and pieces of himself. The only conclusion is that Reagan was an effective leader not in spite of, but because of his double dysfunction. He was able to produce ideological effects by non-ideological means; a global shift in the political direction of the United States by falling apart. His means were affective. Once again: affective, as opposed to emotional. This is not about empathy or emotive identification, or any form of identification for that matter.[11]

Reagan politicized the power of mime. That power is in interruption. A mime decomposes movement, cuts its continuity into a potentially infinite series of submovements punctuated by jerks. At each jerk, at each cut into the movement, the potential is there for the movement to veer off in another direction, to become a different movement. Each jerk suspends the continuity of the movement for just a flash, too quick really to perceive—but decisively enough to suggest a veer. This compresses into the movement under way potential movements that are in some way made present without being actualized. In other words, each jerk is a critical point, a singular point, a bifurcation point. At that point, the mime almost imperceptibly intercalates a flash of virtuality into the actual movement under way. The genius of the mime is also the good fortune of the bad actor. Reagan's gestural idiocy had a mime effect, as did his verbal incoherence, in the register of meaning. He was a communicative jerk. The two levels of interruption, those of linear movement and conventional progressions of meaning, were held together by the one Reagan feature that did, I think, hold positive appeal—the timbre of his voice, that beautifully vibratory voice. Two parallel lines of abstractive suspense resonated together. His voice embodied the resonation. It embodied the abstraction. It was the embodiment of an asignifying intensity doubling his every actual move and phrase, following him like the shadow of a mime. It was the continuity of his discontinuities.[12]

Reagan operationalized the virtual in postmodern politics. Alone, he was nothing approaching an ideologue. He was nothing, an idiocy musically coupled with an incoherence. That's a bit unfair. He was an incipience. He was unqualified and without content. But the incipience that he was was prolonged by technologies of image transmission, and then relayed by apparatuses, such as the family or the church or the school or the chamber of commerce, which in conjunction with the media acted as part of the nervous system of a new and frighteningly reactive body politic. It was on the receiving end that the Reagan incipience was qualified, given content. Receiving apparatuses fulfilled the inhibitory, limitative function. They selected one line of movement, one progression of meaning, to actualize and implant locally. That is why Reagan could be so many things to so many people; that is why the majority of the electorate could disagree with him on every major issue, but still vote for him. Because he was actualized, in their neighborhood, as a movement and a meaning of their selection—or at least selected for them with their acquiescence. He was a man for all inhibitions. It was commonly said that he ruled primarily by projecting an air of confidence. That was the emotional tenor of his political manner, dysfunction notwithstanding. Confidence is the emotional translation of affect as *capturable* life potential; it is a particular emotional expression and becoming-conscious of one's side-perceived sense of vitality. Reagan transmitted vitality, virtuality, tendency, in sickness and interruption. ("I am in control here," cried the general when Reagan was shot. He wasn't, actually.) The actualizations relaying the Reagan incipience varied. But with the exception of the cynical, the aphasic, and the agnosic, they consistently included an overweening feeling of confidence—that of the supposedly sovereign individual within a supposedly great nation at whose helm idiocy and incoherence reigned. In other words, Reagan was many things to many people, but within a general framework of affective jingoism. Confidence is the apotheosis of affective capture. Functionalized and nationalized, it feeds directly into prison construction and neocolonial adventure.

What is of dire interest now, post-Reagan, is the extent to which he contracted into his person operations that might be argued to be endemic to late-capitalist image- and information-based economies. Think of the image/expression-events in which we bathe. Think interruption. Think of the fast cuts of the video clip or the too cool TV commercial. Think of the cuts from TV programming to commercials. Think of the cuts across programming and commercials achievable through zapping. Think of the distractedness of television viewing, the constant cuts from the screen to its immediate surroundings, to the viewing context where other actions are performed in fits and starts as attention flits. Think of the joyously incongruent juxtapositions of surfing the Internet. Think of our bombardment by commercial images off the screen, at every step in our daily rounds. Think of imagistic operation of the consumer object, as turnover time increases as fast as styles can be recycled. Everywhere, the cut, suspense—incipience. Virtuality, perhaps?

Affect holds a key to rethinking postmodern power after ideology. For although ideology is still very much with us, often in the most virulent of forms, it is no longer encompassing. It no longer defines the global mode of functioning of power. It is now one mode of power in a larger field that is not defined, overall, by ideology.[13] This makes it all the more pressing to connect ideology to its real conditions of emergence. For these are now manifest, mimed by men of power. One way of conceptualizing the nonideological means by which ideology is produced might deploy the notions of *induction* and *transduction*—induction being the triggering of a qualification, of a containment, an actualization; and transduction being the transmission of an impulse of virtuality from one actualization to another, and across them all (what Guattari calls transversality). Transduction is the transmission of a force of potential that cannot but be felt, simultaneously doubling, enabling, and ultimately counteracting the limitative selections of apparatuses of actualization and implantation.[14] This amounts to proposing an *analog* theory of image-based power: images as the conveyors of forces of emergence; as vehicles for existential potentialization and transfer. In this, too, there are notable precursors. In particular, Walter Benjamin, whose concept of shock and image bombardment, whose analyses of the unmediated before-after temporality of what he called the "dialectical image," whose fascination with mime and mimicry, whose connecting of tactility to vision, all have much to offer an affective theory of late-capitalist power.[15]

At this point, the impression may have grown that affect is being touted here as a new "theory of everything," as if the whole world could be packed into it. In a way, it can, and is. The affective "atoms" that overfill the jerk of the power-mime are monads, inductive/transductive virtual perspectives fading out in all directions to infinity, separated from one another by dynamic thresholds.[16] They are autonomous, not through closure but through a singular openness. As unbounded "regions" in an equally unbounded affective field, they are in contact with the whole universe of affective potential, as by action at a distance. Thus they have no outside, even though they are differentiated according to which potentials are most apt to be expressed (effectively induced) as their "region" passes into actuality. Their passing into actuality is the key. Affect *is* the whole world: from the precise angle of its differential emergence. How the element of virtuality is construed—whether past or future, inside or outside, transcendent or immanent, sublime or abject, atomized or continuous—is in a way a matter of indifference. It is all of these things, differently in every actual case. Concepts of the virtual in itself are important only to the extent to which they contribute to a pragmatic understanding of emergence, to the extent to which they enable triggerings of change (induce the new). It is the edge of virtual, where it leaks into actual, that counts. For that seeping edge is where potential, actually, is found.

Resistance is manifestly not automatically a part of image reception in late capitalist cultures. But neither can the effect of the mass media and other image and information-based media simply be explained in terms of a lack: a waning of affect, or a decline in belief, or alienation. The mass media are massively potentializing—but that potential is inhibited, and both the emergence of the potential and its limitation are part and parcel of the cultural-political functioning of the media, as connected to other apparatuses. Media transmissions are breaches of indetermination. For them to have any *specific* effect they must be determined to have that effect by apparatuses of actualization and implantation that plug into them and transformatively relay what they give rise to (family, church, school, chamber of commerce, to name but a few). The need actively to actualize media transmission is as true for reactive politics as it is for a politics of resistance, and requires a new understanding of the body in its relation to signification and the ideal or incorporeal. In North America, at least, the far right is far more attuned to the imagistic potential of the postmodern body than the established left, and has exploited that advantage since the mid-1980s. Philosophies of affect, potential, and actualization may aid in finding countertactics.

IV

Last story:

> A man writes a health-care reform bill in his White House. It starts to melt in the media glare. He takes it to the Hill, where it continues to melt. He does not say goodbye.

Although economic indicators show unmistakable signs of recovery, the stock market dips. By way of explanation, TV commentators cite a secondhand feeling. The man's "waffling" on other issues has undermined the public's confidence in him, and it is rebounding on the health-care initiative. The worry is that Clinton is losing his "presidential" feel. What does that have to do with the health of the economy? The prevailing wisdom among the same commentators is that *passage* of the health-care reform bill would harm the economy. It is hard to see why the market didn't go *up* at the news of the "unpresidential" falter of what many "opinion makers" considered a costly social program inconsistent with basically sound economic policy inherited from the previous administration, credited with starting a recovery. However, the question does not even arise because the commentators are operating under the assumption that the stock market registers affective fluctuations in adjoining spheres more directly than properly economic indicators. Are they confused? Not according to certain economic theorists who, when called upon to explain to a nonspecialist audience the ultimate foundation of the capitalist monetary system, answer "faith."[17] And what, in the late-capitalist economy, is the base cause of inflation, according to the same experts? A "mindset," they say, in which feelings about the future become self-fulfilling prophesies capable of reversing "real" conditions (Heilbroner and Thurow 151).

The ability of affect to produce an economic effect more swiftly and surely than economics itself means that affect is itself a real condition, an intrinsic variable of the late-capitalist system, as infrastructural as a factory. Actually, it is beyond infrastructural, it is everywhere, in effect. Its ability to come second hand, to switch domains and produce effects across them all, gives it a metafactorial ubiquity. It is beyond infrastructural. It is transversal.

This fact about affect—this matter-of-factness of affect—needs to be taken seriously into account in cultural and political theory. Don't forget.

Notes

1. The thesis on the waning of affect in Jameson's classic essay on postmodernism ("Cultural Logic") powerfully raised the issue of affect for cultural theory. The most sustained and successful exploration of affect arising from subsequent debates is in Grossberg. The present essay shares many strands with Grossberg's work, including the conviction that affect has become pervasive rather than having waned. Differences with Grossberg will be signaled in subsequent notes.

2. Grossberg slips into an equation between affect and emotion at many points, despite distinguishing them in his definitions. The slippage begins in the definition itself, when affect is defined quantitatively as the strength of an investment and qualitatively as the nature of a concern (82). This is done to avoid the perceived trap of asserting that affect is unformed and unstructured, a move that Grossberg worries makes its analysis impossible. It is argued here that affect is indeed unformed and unstructured, but that it is nevertheless highly organized and effectively analyzable (it is not entirely containable in knowledge, but is analyzable in effect, as effect). The crucial point is that form and structure are not the only conceivable modes of differentiation. Here, affect is seen as prior to or apart from the qualitative, and its opposition with the quantitative, and therefore not fundamentally a matter of investment (if a thermodynamic model applies, it is not classical but quantum and far-from-equilibrium; more on this later). For more on the relation between affect and quality/quantity, see Massumi, "The Bleed."

3. The reference to conventional discourse in Spinoza is to what he calls "universal notions" (classificatory concepts that attribute to things defining structural properties and obey the law of the excluded middle) and "transcendental notions" (teleological concepts explaining a thing by reference to an origin or end in some way contained in its form). See *The Ethics*, book 2, proposition 40, scholium 1 in volume 1 of *The Collected Works of Spinoza*.

4. The retrospective character of attributions of linear causality and logical consistency was analyzed by Henri Bergson under the rubric of the "retrograde movement of truth." See *The Creative Mind*.

5. See, in particular, chapter 2 (an analysis of the chemistry of crystallization). Throughout his work Simondon carries out a far-reaching critique of concepts of form and structure in philosophy and the natural and social sciences.

6. On proprioception and affect, see Massumi, "The Bleed."

7. A connection could be made here with the work of Walter Benjamin on shock and the circulation of images. Susan Buck-Morss (312) quotes from Benjamin's *Passagen-werk* on the "monadological structure" of "dialectical images." This structure is a "force-field" manifesting a nonlinear temporality (a conflict between "fore-history" and "after-history" in direct connection with one another, skipping over the present without which the conflict would nevertheless not take place: "in order for a piece of the past to be touched by present actuality, there must be no connection between them").

8. For a brilliant analysis of affect in terms of intensity, vitality, synaesthesia ("amodal perception"), and nonconscious sense of self, see Stern.

9. Deleuze discusses perception, the brain, and matter in *Cinema 1*, chapters 1 and 3 (in relation to Bergson). Deleuze and Guattari make the connection between the brain and chaos in the conclusion to *What Is Philosophy?*.

10. The main difference between this perspective and that of Lawrence Grossberg is that his approach does not develop a sustainable distinction between implicate and explicate orders (between virtuality and actuality, intension and extension). Although Meaghan Morris does not use the term "affect," her analysis of the function of the TV screen brings her approach to the mass media into close philosophical affinity with the one being developed here. In "Ecstasy and Economics (A Portrait of Paul Keating)," she describes the screen image as triggering a "phase of empowerment" that is also a "passage" and "transport," not between two places but between a place and a nonplace, an "elsewhere": "the screen . . . is not a border between comparable places or spaces . . . What visibly 'exists' there, 'bathed' in glow, is merely a 'what'—a relative pronoun, a bit of language, that *relation* 'your words describe'" (Morris 7–72).

11. On these and other topics, including gory detail of Reagan's crumblings, see Dean and Massumi. The statement that ideology—like every actual structure—is produced by operations that do not occur on its level and do not follow its logic is simply a reminder that it is necessary to integrate implicate order into the account. This is necessary to avoid capture and closure on a plane of signification. It signals the measure of openness onto heterogeneous realities of every ideological structure, however absolutist. It is a gesture for the conceptual enablement of resistance in connection with the real. Ideology is construed here in both the commonsense meaning as a structure of belief, and in the cultural-theoretical sense of an interpellative subject positioning.

12. On mime, see José Gil.

13. For one account of how this larger field functions, see Deleuze, "Post-scriptum," 240–47.

14. The concept of transduction is taken, with modifications, from the work of Gilbert Simondon.

15. In addition to the quotes in Buck-Morss cited in note 7 above, see in particular Benjamin 160–63; see also Michael Taussig 141–48. Bakhtin also develops an analog theory of language and image in which synaesthesia and the infolding of context discussed earlier in this essay figure prominently.

16. Bohm and Hiley (353–54) use a holographic metaphor to express the monadic nature of the "implicate order" as "enfolded" in the explicate order.

17. "Behind [currency], rests the central requirement of faith. Money serves its indispensable purposes as long as we believe in it. It ceases to function the moment we do not" (Heilbroner and Thurow 138).

Works Cited

Bakhtin, Mikhail. "The Problem of Content, Material, and Form in Verbal Art." In *Art and Answerability: Early Philosophical Essays,* edited by Michael Holquist and Vadim Liapunov and translated by Vadim Liapunov. Austin: University of Texas Press, 1990.

Benjamin, Walter. *One Way Street.* London: Verso, 1985.

Bergson, Henri. *The Creative Mind.* Translated by Mabelle L. Audison. New York: Philosophical Library, 1946.

Bohm, David, and B. J. Hiley. *The Undivided Universe.* New York: Routledge, 1993.

Buck-Morss, Susan. "Dream-world of Mass Culture: Walter Benjamin's Theory of Modernity and the Dialectics of Seeing." In *Modernity and the Hegemony of Vision,* edited by Michael Levin. Berkeley: University of California Press, 1993.

———. *Matter and Memory.* Translated N. M. Paul and W. S. Palmer. New York: Zone, 1988.

Dean, Kenneth, and Brian Massumi. *First and Last Emperors: The Absolute State and the Body of the Despot.* New York: Autonomedia/Semiotexte, 1992.

Deleuze, Gilles. *Cinema 1: The Movement-Image.* Translated by Hugh Tomlinson and Barbara Habberjam. Minneapolis: University of Minnesota Press, 1986.

———. *Difference and Repetition.* Translated by Paul Patton. New York: Columbia University Press, 1994.

———. "Post-scriptum sur les sociétés de contrôle." *Pourparlers.* Paris: Minuit, 1990. (Forthcoming in English, New York: Columbia University Press.)

Deleuze, Gilles, and Félix Guattari. *What Is Philosophy?* New York: Columbia University Press, 1993.

Gil, José. *Métamorphoses du corps*. Paris: Editions de la Différence, 1985. In English, *Metamorphoses of the Body*. Translated by Stephen Muecke. Minneapolis: University of Minnesota Press, 1997.

Grossberg, Lawrence. *We Gotta Get Out of This Place: Popular Conservatism and Postmodern Culture*. New York: Routledge, 1992.

Guattari, Félix. *Chaosmose*. Paris: Galilée, 1992. In English, *Chaosmosis: An Ethico-Aesthetic Paradigm*. Translated by Paul Bains. Bloomington: Indiana University Press, 1995.

Heilbroner, Robert and Lester Thurow. *Economics Explained: Everything You Need to Know about How the Economy Works and Where It Is Going*. New York: Simon and Schuster, 1994.

Horgan, John. "Can Science Explain Consciousness?" *Scientific American* (July 1944): 76–77.

Jameson, Fredric. "The Cultural Logic of Late Capitalism." In *Postmodernism: The Cultural Logic of Late Capitalism*. New York: Verso, 1991.

Massumi, Brian. "The Bleed: Where Body Meets Image." In *Rethinking Borders*, edited by John Welchman. London: Macmillan, forthcoming.

Morris, Meaghan. *Ecstasy and Economics: American Essays for John Forbes*. Sydney: Empress Publishing, 1992.

Sacks, Oliver. *The Man Who Mistook His Wife for a Hat*. London: Picador, 1985.

Simondon, Gilbert. *L'individu et sa genèse physico-biologique*. Paris: Presses Universitaires de France, 1964.

Spinoza, Baruch. *The Collected Works of Spinoza*. Edited and translated by Edwin Curley. Princeton: Princeton University Press, 1985.

Stern, Daniel. *The Interpersonal World of the Infant: A View from Psychoanalysis and Developmental Psychology*. New York: Basic Books, 1985.

Sturm, Hertha. *Emotional Effects of Media: The Work of Hertha Sturm*. Edited by Gertrude Joch Robinson. Montreal: McGill University Graduate Program in Communications, 1987.

Taussig, Michael. "Tactility and Vision." In *The Nervous System*. New York: Routledge, 1992.

Contributors

Drucilla Cornell is professor of law at Rutgers University, Newark. Her books include *The Imaginary Domain: Abortion, Pornography, and Sexual Harassment*; *Beyond Accommodation: Ethical Feminism, Deconstruction, and the Law*; *The Philosophy of the Limit*; and *Transformations: Recollective Imagination and Sexual Difference*.

Jonathan Elmer is professor of English at Indiana University. He is the author of *Edgar Allan Poe and the Imagination of Mass Culture*.

N. Katherine Hayles is professor of English at the University of California, Los Angeles. She is the author of *Chaos and Order: Complex Dynamics in Literature and Science*; *Chaos Bound: Orderly Disorder in Contemporary Literature and Science*; *The Cosmic Web: Scientific Field Models and Literary Strategies in the Twentieth Century*; and *How We Became Posthuman: Virtual Bodies in Cybernetics, Literature, and Informatics*. She is the co-editor, with Mark Poster and Samuel Weber, of the book series Electronic Mediations, published by the University of Minnesota Press.

Peter Uwe Hohendahl is the Jacob Gould Schurman Professor of German and comparative literature at Cornell University. His books include *The Institution of Criticism*; *Reappraisals: Shifting Alignments in Postwar Critical Theory*; *Prismatic Thought: Theodor W. Adorno;* and the edited collection *A History of German Literary Criticism, 1730–1980*.

Eva Knodt is the author of *Negative Philosophie und dialogische Kritik* and numerous articles on eighteenth-century Germany. She has also written on Luhmann and Habermas (most notably, the introduction to Luhmann's *Social Systems*) and coedited a special issue of *New German Critique*.

Marjorie Levinson is professor of English at the University of Michigan. She is the author of *The Romantic Fragment Poem*; *Wordsworth's Great Period Poems*; and *Keat's Life of Allegory;* and the editor of *Rethinking Historicism.*

Niklas Luhmann was, until his death in 1998, the foremost proponent of a general theory of functionally differentiated, self-reproducing social systems. Among his books that have been translated into English are *Social Systems*; *Essays on Self-Reference*; *Ecological Communication*; *Political Theory of the Welfare State; Love as Passion: The Codification of Intimacy*; and *The Differentiation of Society.*

Brian Massumi is professor of English at SUNY-Albany. He is the author of *User's Guide to Capitalism and Schizophrenia: Deviations from Deleuze and Guattari* and *First and Last Emperors: The Absolute State and the Body of the Despot* (with Kenneth Dean). He is the coeditor of the University of Minnesota Press book series Theory Out of Bounds, and he has translated numerous books, including *A Thousand Plateaus* by Gilles Deleuze and Félix Guattari, published by the University of Minnesota Press.

William Rasch is professor of Germanic studies at Indiana University. He has published extensively on Niklas Luhmann and systems theory, Carl Schmitt, and the late eighteenth century.

Cary Wolfe is professor of English at SUNY-Albany. He is the author of *Critical Environments: Postmodern Theory and the Pragmatics of the "Outside"* (Minnesota, 1998) and *The Limits of American Literary Ideology in Pound and Emerson.*

Permissions

The University of Minnesota Press gratefully acknowledges permission to reprint the following essays in this volume.

William Rasch, "The Limit of Modernity: Luhmann and Lyotard on Exclusion," originally appeared in *Soziale Systeme* 3, no. 2 (1997): 257–69. Reprinted here by the generous permission of the editors.

Cary Wolfe, "Making Contingency Safe for Liberalism: The Pragmatics of Episte mology in Rorty and Luhmann," originally appeared in *New German Critique* 61 (winter 1994). Reprinted here with the kind permission of Telos Press, Ltd.

The following essays were originally published in *Cultural Critique* and are reprinted here by permission of Oxford University Press, for which the editors of this volume would like to express their thanks.

Reprinted from *Cultural Critique* 30 (spring 1995):

William Rasch and Cary Wolfe, "Introduction," pp. 5–13. (The introduction to this book volume has been expanded and modified.)

Cary Wolfe, "In Search of Posthumanist Theory: The Second-Order Cybernetics of Maturana and Varela," pp. 33–70.

N. Katherine Hayles, "Making the Cut: The Interplay of Narrative and System, or What Systems Theory Can't See," pp. 71–100.

Jonathan Elmer, "Blinded Me with Science: Motifs of Observation and Tempo-rality in Lacan and Luhmann," pp. 101–36.

Niklas Luhmann, "Why Does Society Describe Itself as Postmodern?," pp. 171–86.

Peter Uwe Hohendahl, "No Exit? (Response to Luhmann)," pp. 187–92.

William Rasch, "Immanent Systems, Transcendental Temptations, and the Limits of Ethics," pp. 193–221.

Drucilla Cornell, "Rethinking the Beyond within the Real (Response to Rasch)," pp. 223–34.

Reprinted from *Cultural Critique* **31 (fall 1995):**

"Theory of a Different Order: A Conversation with Katherine Hayles and Niklas Luhmann," pp. 7–36.

Brian Massumi, "The Autonomy of Affect," pp. 83–109.

Marjorie Levinson, "Pre- and Post-Dialectical Materialisms: Modeling Praxis without Subjects and Objects," pp. 111–27.

Index